SO-AKE-781

Fun with Mathematics

Fun with Mathematics

By PHILIP HEAFFORD

BELL PUBLISHING COMPANY
NEW YORK

Copyright © MCMLIX by Emerson Books, Inc.
All rights reserved.

This 1983 edition is published by Bell Publishing Company,
distributed by Crown Publishers, Inc., by arrangement with
Emerson Books, Inc.

Originally published as *The Math Entertainer*

Manufactured in the United States of America

Library of Congress Cataloging in Publication Data

Heafford, Philip Ernest.
  Fun with mathematics.
  Reprint. Originally published: The math
entertainer. New York : Emerson Books, 1959.
  1. Mathematical recreations.   I. Title.
QA95.H37   1982   793.7′4   82-12834

ISBN: 0-517-393972

h g f e d c b a

To
all those who
love to solve
a problem

CONTENTS

[7]

INTRODUCTION

The author and a friend, Henry Babb, were recently on a long touring holiday in Europe. While they were on this vacation, the conversation drifted over to mathematics and in particular to the secret of success in winning over those who say:

> Multiplication is vexation,
> Division's twice as bad,
> The rule of three perplexes me,
> And fractions drive me mad.

There is no doubt that many children suffer from gaps in their knowledge of mathematics, caused perhaps by insufficient care in the planning of lessons, and this is likely to result in a distaste for and even hostility to mathematics. The author knows only too well that, in this subject of study more than in any other, there is no short cut and that the learner must proceed methodically step by step. He believes that there is no necessity for any "hostility" to mathematics. Indeed, he thinks that mathematics can be entertainment and can lead to much joy and satisfaction, provided that the gradient of achievement is properly adjusted to the ability of the individual.

It is certain that in mathematics nothing succeeds like success. But, of course, some effort is required if anyone is to be moderately successful in the various branches of this subject, just as it is in learning to swim, to dance, or to hit a golf ball six times out of six. When once you are beyond the five-finger exercises in mathematics, the fun can begin. This has been proved by hundreds of men and women, boys and girls.

[9]

The conversation of the author and his friend was, in part, a pedagogically serious discussion. They agreed that one of the secrets of success in teaching students is to make lessons and lectures interesting by historical asides and unusual problems. Out of their agreement this quiz book was born. The various sets of questions were quickly drafted, and many of the answers were scribbled down without reference to books. The author has had the assistance of friends in other countries who have been able to tap several sources of information unfamiliar to himself.

Perhaps this book will bring to some, whose memories of arithmetic, algebra, geometry, and trigonometry suggest toil and possibly tears, a different and a happier prospect. The author and his friend have themselves found in this subject an inspiration and a delight—in fact, fun. They say, "If everybody has to use mathematics in some form or other, it is only fair that everybody should get some fun out of it." They hope that the reader will soon be able to join hands with the Major-General and sing from *The Pirates of Penzance*:

I'm very well acquainted too with matters mathematical,
I understand equations, both the simple and quadratical,
About binomial theorem I'm teeming with a lot of news—
With many cheerful facts about the square of the hypotenuse.

The author hopes to interest an audience that already has some interest in mathematics, and to attract an audience that may be stimulated to take an interest. Hence some of the quizzes are a very mixed "bag," containing questions of much diversity in subject matter and a wide range of difficulty.

The author hopes that he has provided something for everybody.

Oxford   P. H.

QUESTIONS

Do these numbers ring a bell? For instance, the number 365 would mean only one thing to me, and that is the number of days in a year. Ask someone to test you with this quiz. Six seconds for each question. How many can you get right in the time limit of two minutes for all the questions?

1. 1,760	11. .4771
2. 2,000	12. .4971
3. 4,840	13. 1.6
4. 640	14. 1.414
5. 1.732	15. 1,728
6. 2.54	16. 3-4-5
7. 3.1416 . . .	17. 6,080
8. 366	18. 62½
9. .3010	19. 90
10. 1492	20. 88

 Answers on page 67

THE PRINTER'S NIGHTMARE

Before the days of the typewriter, the printer's lot was not always a happy one. Imagine how difficult it must have been for the unfortunate printer trying to set up the type for an arithmetic book when the hand-written manuscript was illegible. One printer overcame this difficulty by putting "stars" for the figures he could not decipher. See if you could have helped him by finding out what the figures really are.

1. Addition:

★22★	113
1★★1	6★4
———	14★
3489	★26
———	————
	★410
	————

2. Subtraction:

4★★2	6★35
★35★	★82★
———	———
121	4★7
———	———

3. Multiplication:

★7	★★★7
★★	★★★
———	————
★★★	★★★★6
★★5	★★203
———	★37★★
★★91	————
———	★★★★★★★
	————

4. Equations:

$5x - 5 = ★x - 3$ $x^2 - 4x = ★★$

∴ $x = 2$ ∴ $x = 7$ or ★

[14]   *Answers on page 68*

There are many recognized mathematical symbols. You will know nearly all of them, but it is interesting to know who introduced them. We take so much for granted these days, and it is only when we stop to think that we appreciate much we so often dismiss as commonplace. So when you so readily answer the questions asked here, try to name the person who introduced the symbol you give for each answer. (This cannot be done in every case.) What is the mathematical symbol for ....?

1. equals

2. multiplied by

3. square root of

4. varies as

5. greater than

6. infinity

7. "as," in "*a* is to *b* as *c* is to *d*"

8. the number of combinations of *n* things taken *r* at a time

9. the ratio of the circumference to the diameter for any circle

10. *a* times *a* times *a*

Can you solve these problems?

1. If five girls pack five boxes of flowers in five minutes, how many girls are required to pack fifty boxes in fifty minutes?

2. A boy has a long cardboard strip 1 inch wide and 48 inches long. It is marked at 1-inch intervals so that he can cut off a series of square inches. If the boy takes one second for each cut, how long will it take to cut the 48 square inches?

3. To move a safe, two cylindrical steel bars 7 inches in diameter are used as rollers. How far will the safe have moved forward when the rollers have made one revolution?

4. A town in India has a population of 20,000 people. 5 per cent of them are one-legged, and half the others go barefoot. How many sandals are worn in the town?

5. Without introducing + signs, arrange six "nines" in such a way that they add up to 100.

6. What is there peculiar about the left-hand side of $50\frac{1}{2} + 49\frac{38}{76} = 100$?

7. A fish had a tail as long as its head plus a quarter the length of its body. Its body was three-quarters of its total length. Its head was 4 inches long. What was the length of the fish?

QUIZ NO. 5   ARE YOU AT HOME IN ROME?

For most of the answers to this quiz you will have to know the Roman figures. As they had no zero to give their numbers a "place value," it must have been very awkward when it came to multiplication!

1. What aid was used by the Romans to help with calculations?

2. The following is cut on a famous monument: MDCCLXXVI. What year does this represent?

3. Write 1789 in Roman figures.

4. What is the largest number you can write using these Roman numerals once each, I,C,X,V,L?

5. What is the smallest number you can write using the same Roman numerals once each, I,C,X,V,L?

6. Without changing to our Hindu-Arabic notation, find the value of CXVI + XIII + VI + CCLXV.

7. What Roman numbers of two integers between one and twenty become larger when the left-hand integer is omitted?

8. Was a "groma" used by the Roman merchant, surveyor, cook, or sailor?

1. During a vacation it rained on thirteen days, but when it rained in the morning the afternoon was fine, and every rainy afternoon was preceded by a fine morning. There were eleven fine mornings and twelve fine afternoons. How long was the vacation?

2. At what time between 7 and 8 o'clock will the two hands of a clock be in a straight line?

3. If $11^3 = 1,331$ and $12^3 = 1,728$, what is the cube root of the perfect cube 1,442,897?

4. A bottle of cider costs 25 cents. The cider cost 15 cents more than the bottle. How much should you receive on returning the bottle?

5. The lengths of the sides of a right-angled triangle measure an exact number of feet. If the hypotenuse is 1 foot longer than the base and the perpendicular is 9 feet long, how long are the sides?

6. A spruce tree when planted was 3 feet high and it grew by an equal number of feet each year. At the end of the seventh year, it was one-ninth taller than at the end of the sixth year. How tall was the tree at the end of the twelfth year?

7. Without doing the actual division state whether 13,972,536 is exactly divisible by 8.

8. A cement mixture costs $33 a ton. It is composed of Grade A cement at $36 a ton and Grade B cement at $24 a ton. How were these two cements mixed?

ACROSS

1. The number of cubic inches in 1 cubic foot.

3. The number of yards in 1,188 inches.

5. The arithmetical mean of 2 and 50.

6. The value of g in cm. per sec. per sec.

7. The number of feet in 1 mile.

10. The simple interest on $18,900 for four years at 2½ per cent.

12. The number of ways in which four boys and four girls can sit at a round table so that no two boys sit together.

14. The angle subtended at the center by the side of a regular octagon inscribed in a circle.

16. The diagonal of a rectangular plot 20 yards long and 15 yards wide.

17. The sixth term of the series 6, 18, 54, ——.

DOWN

1. The number of sides in a duodecagon.

2. 1,122 ft. per sec. expressed m.p.h.

3. The size of an angle opposite an angle of 142° in a cyclic quadrilateral.

4. The value of π, correct to three places of decimals, multiplied by 1,000.

6. The smallest number divisible by 11 and greater than 9,000.

8. The area to the nearest square foot of a circular track 10 feet wide and with an inner circumference of 250 feet. ($\pi = 3\frac{1}{7}$.)

9. The third leap year in the nineteenth century.

11. The bearing understood by an air navigator equivalent to the sailor's direction N.E.

13. The angle which the graph of $y = x$ makes with the x — axis.

15. The length of the hypotenuse of a 30°, 60°, 90° triangle if the side opposite the 30° is 29 feet.

Answers on pages 79-80

QUIZ NO. 8   *THE TRIANGLE TEST*

A triangle is a geometrical figure bounded by three straight lines and having three angles. Such a definition may be correct, but it gives one the idea that a triangle is a decidedly uninteresting figure. There are many different kinds of triangles and each one has its own interesting peculiarities. From the information given, can you state the names of these triangles?

1. I am readily suggested when you look at a trillium.

2. I appear when a man stands on level ground with his legs straight and his feet slightly apart.

3. I have a special name derived from a Greek word meaning "uneven."

4. I am formed by joining the feet of the perpendiculars from the vertices of a triangle to the opposite sides.

5. The sum of the squares on two of my sides equals the square constructed on my third side.

6. There are at least two of us. We find that our corresponding angles are equal and our sides are proportional.

7. The sides and the diagonals of a quadrilateral are used to construct me.

8. My sides are not straight lines and the sum of my angles is greater than 180°.

9. I have gained the title "pons asinorum" for a certain proposition in Euclid.

10. I am connected with the stars and the zenith.

[20]   *Answers on pages 81-82*

There are a large number of mathematical terms that are included in expressions in common use. How frequently we hear "equal rights," "shooting a line," "hot rod," "100-percent effort," "integral part," "vicious circle." You will be able to find many more if you listen carefully. In the following, find the mathematical term that is a part, or the whole, of an everyday expression suggested by:

1. The area over which anything exerts influence.

2. Having equal scores when playing golf.

3. The hour at which an operation is timed to begin.

4. The Great —— between Atlantic and Pacific.

5. A phrase implying excess and having no relation in size, amount, etc.

6. The modern measuring unit of intelligence.

7. A famous military building.

8. One who sets forth in words, expounds, or interprets.

9. The rejection of a person proposed for some office.

10. A judge recapitulates the evidence at the end of a case.

Answers on pages 83-84

In this quiz you will find ten anagrams. Each question is an anagram of the name of a well-known mathematician. To help you, a sentence is added which refers to the person named in the anagram.

1. DIME SEARCH. "If I had a fulcrum I could move the world."

2. SCARED SET. He did not belong to this set, for he was a soldier before he was attracted to mathematics.

3. MENU SALE. He was one of the few of the first century A.D. who did original mathematical work of any ability.

4. A DACRON. He could not have "obtained" a dacron shirt, but he seems to have "acquired" a contemporary's solution of a cubic equation.

5. PANIER. He was not a Frenchman, and his "bones" were not kept in a basket.

6. WAS ILL. He had much to do with the founding of the Royal Society in England.

7. RED TOUGH. Yet he was delighted to hear of the king's return, and even invented a "rule."

8. ALL PACE. He discovered a famous differential equation.

9. RULE E. A Swiss mathematician, and a certain constant has been given his name.

10. ROME DIVE. He was chosen to decide the controversy over the discovery of the calculus.

1. There are three books, each one inch thick. They stand side by side in order—Volumes I, II, and III. A bookworm starts outside the front cover of Volume I and eats its way through to the outside of the back cover of Volume III. If the worm travels in a straight line, how far does it travel?

2. A man built a cubical house with ordinary windows in all the upright walls. He found whenever he looked out of a window that he was looking south. Where did he build his house?

3. A merchant has two large barrels. The smaller barrel holds 336 liters but is only five-sixths full of wine. He empties this wine into the other barrel and finds that the wine fills only four-ninths of it. How much wine would the larger barrel hold when full?

4. What three curves are produced by making sections of a right circular cone in directions other than parallel to the base?

5. Two men play a card game and the stake is one penny a game. At the end one has won three games and the other has won three pennies. How many games did they play?

6. A number consists of three digits, 9, 5, and another. If these digits are reversed and then subtracted from the original number, an answer will be obtained consisting of the same digits arranged in a different order still. What is that other digit?

There are many interesting things to know about circles. One or two of these may make you think!

1. Complete the following verse:
    "Tweedledum and Tweedledee
    Around the circle is π times ——.
    But if the area is declared
    Think of the formula ——."

2. What two words derived from the Latin for "a bow" and "a string" are used in the geometry of the circle?

3. What is meant by "squaring the circle"?

4. What important ratio was known until 1736 as "*c*" or "*p*"?

5. The minute hand of a clock is 7 inches long. What distance does the tip of the hand move in 22 minutes?

6. What path is traced out by a mark on the tire of a bicycle wheel as the cycle is ridden in a straight line on a level surface?

7. In reference to question 6, what distance does this mark actually move through as the wheel revolves once?

8. Construct semicircles on the three sides of an isosceles right-angled triangle. You will form two lunes. Show that each lune equals half the area of the triangle.

[24]   *Answers on pages 90-91*

QUIZ NO. 13 SOME OLD AND SOME NEW

1. Find a quantity such that the sum of it and one-seventh of it shall equal nineteen.

2. How many guests were present at a Chinese party if every two used a dish for rice between them, every three a dish for broth, every four a dish for meat, and there were 65 dishes altogether.

3. A retired colonel lived a quarter of his life as a boy, one-fifth as a young man, one-third as a man with responsibilities, and thirteen years on pension. How old was he when he died?

4. The fat men in a club outnumber the thin men by sixteen. Seven times the number of fat men exceeds nine times the number of thin men by thirty-two. Find the number of fat and thin men in the club.

5. An explorer grew a beard during his travels. At the end of his journeys, he found that double the length of his whiskers added to its square plus twenty exactly equalled the number of days he had been away. If he had measured the length of his beard in centimeters, and if he had been away 140 days, how long was his beard at the end of his travels?

6. A cathedral tower 200 feet high is 250 feet from a church tower 150 feet high. On the top of each tower is a pigeon. The two pigeons fly off at the same time and at the same speed directly to some grain on the level straight road between the towers. The pigeons reach the grain at the same instant. How far is the grain from the foot of the cathedral tower?

Answers on pages 92-93

QUIZ NO. 14   *SEE AND PERCEIVE*

Lots of people see things every day and yet they don't see them. Because things are familiar they often go unnoticed. Do you know how many stairs you climb to the second floor in your house? Can you state what particular geometrical shape is associated with . . . ?

1. the cell in a honeycomb?

2. a roof truss or a braced shelf bracket?

3. a Norman arch in church architecture?

4. an open pantograph?

5. a Maltese cross?

6. the cross-section of the stem of a plant of the Labiatae family?

7. the lines joining the outermost tips of the corolla of a campanula?

8. a "diamond" on a playing card?

9. the constellation Pegasus?

10. the three stars Betelgeuse, Sirius, and Procyon?

WHY MAKE IT A
DIFFICULTY?

It is really most interesting to read about our system of weights and measures, but any detached onlooker must be highly amused at the way we tenaciously make things difficult for ourselves. Why do we not adopt the metric system? Is there an answer to this stupidity? Shall we continue to torture ourselves for ever? Do you know . . . ?

1. the name of the Roman pound?

2. the Latin word meaning "a twelfth part"?

3. where the Troy of troy weight is situated, or is it the name of a man?

4. the measure of weight used to weigh diamonds?

5. the body measurement now standardized as four inches?

6. what unit is derived from the Latin word meaning "a field"?

7. what other name is given to a rod or perch?

8. the name given to the twenty-fourth part of an ounce by a pharmaceutical chemist?

9. the king who is supposed to be connected with our standard measurement of a yard?

10. a particular "weight" that is true to its name in the United States of America and not in Great Britain?

[27]   *Answers on pages 96-97*

Merely because a statement appears in print it is not necessarily accurate! How often one hears the remark, "I'll show it to you in black and white," as if that is sufficient to decide whether something is true. A mathematician must always be accurate. Are the following statements true or false?

1. The pentagram of Pythagoras is formed by drawing all the diagonals of a regular pentagon.

2. Archimedes was the originator of the well-known puzzle of Achilles and the tortoise.

3. 1:05 p.m. is sometimes written as 1305 hours.

4. The curve in which a uniform cable hangs when suspended from two fixed points is a parabola.

5. A pantograph is a mechanical device for drawing figures similar to given figures.

6. A histogram is a hundred kilograms, and this standard unit is kept at the International Bureau of Weights and Measures at Sèvres, near Paris.

7. A cantilever beam is a beam supported at one end only and extending horizontally.

8. A parameter is an independent variable in terms of which the co-ordinates of a variable point may be expressed.

For purposes of identification certain lines have been given special names, e.g. a tangent, an arc, and a radius. You have to name the line referred to in each of these questions. I . . .

1. join the vertex of a triangle to the mid-point of the opposite side.

2. was said to be the shortest distance between two points.

3. subtend a right angle at the circumference of a circle.

4. am the line so drawn in a circle that the angle between me and a certain tangent is equal to the angle in the alternate segment.

5. "touch" a hyperbola at an infinite distance.

6. cut a circle in two points.

7. join all the points of the same latitude on the earth.

8. am the locus of a point from which the tangents drawn to two given circles are equal.

9. am the essential straight line which, together with the special point or focus, enables points on an ellipse or parabola to be determined.

10. pass through the feet of the perpendiculars drawn to the three sides of a triangle from any point on the circumcircle of the triangle.

QUIZ NO. 18 *LIKE BUT UNLIKE*

There is a suffix "-oid" which is usually found at the end of Greek roots. This suffix means "having the form of," "like," or "resembling." For instance, the earth is said to be an oblate spheroid, which means that it is like a sphere which is flattened (oblate) at the poles. In mathematics you will sometimes find a word ending in "-oid," and in this quiz you have to give a one-word answer using this suffix.

1. The surface made by revolving a parabola about its principal axis.

2. Nearly like the surface of a football.

3. The medians of a triangle have this point in common.

4. The curve traced out by a mark on the edge of a dime when rolled around the inside edge of a fixed tea cup or a cylindrical can.

5. The curve traced out by a mark on the edge of a dime when rolled around the edge of a silver dollar.

6. It is an "ivy-shaped" curve and was discovered by Diocles.

A MATHEMATICAL MIXTURE

This is a mixed bag of questions. Some are easy and some are hard. There is no connection between them whatsoever. Get busy as the proverbial bee and count how many you can answer correctly. Perfect marks will qualify you for the award of the Pythagorean star which you can draw for yourself. Do you know . . . ?

1. the number of barleycorns in an inch?

2. the instrument used by Sir Francis Drake to find the altitude of the sun and hence the time?

3. the instrument used in the sixteenth century to tell the time at night by observing the constellation Ursa Major?

4. the name of the mathematician who first proved

$$\Delta = \sqrt{s(s - a)(s - b)(s - c)} \, ?$$

5. the name given to the figure like a five-pointed star often used in the Middle Ages to frighten away witches?

6. what "meter" is used to measure the area contained by a closed plane curve?

7. the name of the solid formed by cutting a pyramid or a cone by two parallel planes?

8. to what use Simpson's rule is put?

9. the common name for a regular hexahedron?

10. how long a clock will take to strike "twelve" if it takes five seconds to strike "six"?

Answers on pages 105-106

In the following sentences will be found one, two, or three words that form the anagram of the name of a mathematician.

1. Those attending his classes were either "listeners" or "mathematicians," but he was not heard to say "graph" as we understand the term.

2. He formed a school called the "Academy," and his many students always found in him a pal to help them.

3. "Number work cannot be taught by the rod alone" is a fact recognized by all teachers since he introduced Arabic numerals into Europe.

4. He observed a swinging lamp and was not interested in whether a gel oil was a contradiction in terms.

5. There is no doubt that many a clasp went into the construction of the arithmetical machine that he invented.

6. Men like him would rob war of its sting, for he recognized and frankly acknowledged the superiority of his pupil.

7. Some of his ideas were not new, but there is no doubt that he was without equal.

8. The geometrical interpretation of complex numbers on a blackboard can be rubbed off with a rag and some "elbow grease"!

1. A dear old Grandpa named Lunn
   Is twice as old as his son.
      Twenty-five years ago
      Their age ratio
   Strange enough was three to one.

   When does Grandpa celebrate his centenary?

2. Said a certain young lady called Gwen
   Of her tally of smitten young men,
      "One less and three more
      Divided by four
   Together give one more than ten."

   How many boy friends had she?

3. There was a young fellow named Clive,
   His bees numbered ten to the power five.
      The daughters to each son
      Were as nineteen to one,
   A truly remarkable hive!

   How many sons (drones) were in the hive?

4. A team's opening batter named Nero
   Squared his number of hits, the hero!
      After subtracting his score,
      He took off ten and two more,
   And the final result was a "zero."

   How many hits did Nero make?

5. Some freshmen from Trinity Hall
   Played hockey with a wonderful ball;
      They found that two times its weight,
      Plus weight squared, minus eight,
   Gave "nothing" in ounces at all.

   What was the weight of the ball?

[33]   *Answers on pages 110-111*

QUIZ NO. 22 **BREVITY IN**
MATHEMATICS

The mathematician frequently uses abbreviations in his work. For the word "logarithms" he uses the shortened term "logs," and for "simple harmonic motion" he uses the initial letters of these words and writes "S.H.M." What abbreviation does he use for . . . ?

1. "which was to be proved or demonstrated"?

2. the cosine of the angle θ?

3. an expression which depends for its value on the value you give to x?

4. the integration of $16x^3$ with respect to x?

5. the smallest number which is exactly divisible by two or more numbers?

6. the hyperbolic sine of x?

7. the square root of -1?

8. the greatest number which will divide exactly into two or more numbers?

9. the derivative of y with respect to x?

10. the eccentricity of conics?

It is always fun discovering the derivation of a word. There is no doubt that even a little acquaintance with Latin and Greek enables one to appreciate why some subjects or figures are named as they are. You are not given in this quiz the Greek, Latin, or Arabic "source words" but translations of them. You have to state what the resulting word is in English. Most of the answers will consist of mathematical words that are nouns and most of these will be the names of various branches of mathematics. The derivation is from . . .

1. two Greek words, meaning "a star" and "to arrange."

2. a Greek word, meaning "the art of counting."

3. three Greek words, suggesting "the measurement of a triangle."

4. two Greek words, meaning "the earth" and "to measure."

5. the title of an Arabic book. Various translations of the title are possible, and of these "the reunion of broken parts" is probably the simplest.

6. a Greek word that may be expressed as "relating to learning."

7. the Greek meaning of "causing to stop or stand."

8. a Latin word, meaning "a little stone."

Necessity is said to be the mother of invention. This is certainly true as far as the instruments and apparatus of mathematics are concerned. All of you will know something about the instruments whose names are required as answers to the following. Name the instrument . . .

1. known in your nursery days as a bead frame.

2. called for simplicity "a water clock."

3. used in the Middle Ages in England to keep accounts, and looking like a notched piece of wood.

4. made of thin wood (or plastic) with edges shaped to help curve drawing.

5. used on board ship for astronomical observations.

6. handy for surveying and measures angles in a vertical and horizontal plane.

7. which when first invented utilized two Gunter's scales.

8. invented by Blaise Pascal, and is now a boon to many clerks.

9. which might also be an apt description of a school principal when everything has gone wrong.

10. employed to measure the thickness of a wire to within 0.001 cm.

1. What is the name of the small metal frame with a glass or plastic front on which is a fine black line? It is used to facilitate the reading of a slide rule.

2. What is constructed in the same ratio as the following numbers? 24 : 27 : 30 : 32 : 36 : 40 : 45 : 48.

3. Two half dollars, *H* and *D*, are touching one another. *H* is rolled around *D* without slipping. How many times will *H* revolve around its own center by the time it is back in its original position?

4. What curve has been called the "Helen of Geometers"?

5. How can you plant ten tulips in ten straight rows with three tulips in each row?

6. The diameter of a long-playing record is 12 inches. The unused center has a diameter of 4 inches and there is a smooth outer edge 1 inch wide around the recording. If there are 91 grooves to the inch, how far does the needle move during the actual playing of the recording?

7. Two men, Mr. Henry and Mr. Phillips, are appointed to similar positions. One elects to receive a beginning salary of $3,000 per year with increases of $600 each year, and the second, Mr. Phillips, chooses a beginning salary of $1,500 per half-year and an increase of $300 every six months. Which person is better paid?

Below you will find some problems that were common in arithmetic textbooks fifty years ago. So often Mr. A, Mr. B, and Mr. C appeared, and the unfortunate Mr. C seemed to be the loser, or the person who got the worst of everything! If ever a single person deserves lasting credit from authors it is surely Mr. C. There are no rivals for that honor! Turn the clock back fifty years and solve the following:

1. A field is owned by three people; *A* has three fifths of it, and *B* has twice as much as *C*. What fraction of the field belongs to *C*?

2. In a mile race *A* beats *B* by 20 yards, and he beats *C* by 40 yards. By how much could *B* beat *C* in a mile race?

3. *A* and *B* can do a piece of work in ten days; *A* and *C* can do it in twelve days; *B* and *C* can do it in twenty days. How long will *C* take to do the work alone?

4. During a game of billiards *A* can give *B* 10 points in 50, and *B* can give *C* 10 points in 50. How many points in 50 can *A* give *C* to make an even game?

5. *A*, *B*, and *C* form a partnership. *A* furnishes $1,875, *B* furnishes $1,500, and *C* $1,250 capital. The partnership makes a profit of $1,850 in the first year. What should *C* take as his share of the profit?

6. Pipes *A* and *B* can fill a tank in two hours and three hours respectively. Pipe *C* can empty it in five hours. If all be turned on when the tank is empty, how long will it take to fill?

[38] *Answers on pages 120-121*

LETTERS FOR
NUMERALS

*Some simple sums were prepared using the numerals 0 to 9.
Then all the numerals were changed to letters. You have to
discover the code which was used for the change. You can
do this if you look carefully for every possible clue. There is
no need to guess. Work these clues methodically, trying each
possibility one after the other. There is only one solution to
each sum. The code has been changed for each sum. Don't
peep at the answers until you have finished and checked
your calculation, because the knowledge of one single
change will make it too easy and spoil your fun.*

1. Addition

```
    X X X X
    Y Y Y Y
    Z Z Z Z
  _____
    Y X X X Z
  _____
```

3. Division

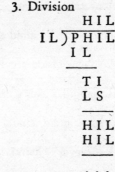

```
              H I L
         _____
    I L ) P H I L
           I L
         _____
           T I
           L S
         _____
           H I L
           H I L
         _____
           . . .
```

2. Multiplication

```
      P N X
        N X
    _____
      R N X
    N X S
    _____
    Z P N X
    _____
```

4. Division

```
              Y F Y
          _____
    A Y ) N E L L Y
           N L Y
         _____
           P P L
           P N H
         _____
           N L Y
           N L Y
         _____
           . . .
```

Answers on pages 122-124

WHO COULD HAVE THIS ON HIS FAMILY CREST?

The crests of towns, cities, and families all obey certain well-defined laws of heraldry. Sometimes these crests are divided into two, and sometimes into four, sections. On the family crest belonging to Sir Isaac Newton there might be an apple tree in one section and a spectrum in another. If a mathematician is noted for some special achievement, it would be reasonable to include something connected with this on his family crest. Who might have the following on theirs?

1. A mural quadrant.

2. The Great Pyramid and its shadow.

3. A sphere inside a cylinder.

4. Rectangular axes marked XOX' and YOY'.

5. A cyclic quadrilateral with the two diagonals drawn.

6. A surveyor's chain with 66 feet marked above it.

7. An ellipse with the sun at one focus.

8. The leaning tower of Pisa.

QUIZ NO. 29   ARCHES

The application of geometry to architectural drawings is obvious, and a mathematician can be an interesting companion on a sight-seeing tour. One thrilled a group of students when he showed them how a particular arch in an old church could be drawn readily with the aid of a pair of compasses and a ruler. Can you spot the arch that is suggested by the following statements?

1. It sounds like an exclamation, but is an arch of **two** double curves that rise to a point.

2. It is common in Early English churches, and is also the name of a surgical instrument.

3. Very good food is suggested! The Mohammedan race inhabiting Northwest Africa never used this arch.

4. This arch is not connected with heraldry, but is **used** to support a flight of solid steps.

5. Obviously very much connected with a certain kind of triangle.

6. The commonest brick arch in house construction.

7. A rounded arch of more than a semicircle.

8. Most likely to be found in spacious buildings constructed in England between 1485 and 1546.

9. I am semicircular in shape and often have a chevron ornamentation.

The answer to each question in this quiz may be expressed as a single letter of the alphabet. Rearrange these letters to form the name of a well-known college.

1. What is the initial letter of a line which touches a circle?

2. Counting through the alphabet thus: $1 = a$, $2 = b$, $3 = c$, etc., what is the letter which represents a score?

3. What is the letter which looks like the answer to the multiplication 0×123?

4. What letter represents the time rate of change of velocity?

5. What is the abbreviation for a certain unit of power? — P.?

6. What is the abbreviation for a standard unit of length in the metric system?

7. What is the initial letter of a polygon having ten sides?

8. What is the second letter of the name given to an equation of the type $ax^2 + bx + c = 0$?

9. What is missing in this formula for the calculation of simple interest: $\dfrac{P \times \quad \times T}{100}$?

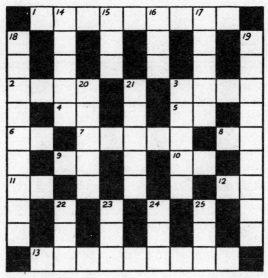

ACROSS

1. He is otherwise known as Leonardo of Pisa.

2. Pascal found work on the cycloid a —— for toothache.

3. Half of brackets.

4. Take nothing from you.

5. The first two letters of the adjective describing the numbers 5, 7, 9, 11, 13, 15, . . .

6. The abbreviation for a coin once used in India.

7. A severely beheaded astronomer of the 17th century.

8. A degree in mathematics.

9. It is, that is, ——.

10. An abbreviation for a unit of weight.

11. The initials for the author of *The Whetstone of Witte.*

12. Abbreviation for 1,000 grams.

13. Il définit les lois des leviers et du centre de gravité des corps.

DOWN

3. He wrote *Finite Differences.*

14. His name suggests "tusks."

15. Unit of electrical resistance.

16. A part of a curve.

17. The straight line joining two points on a curve.

18. An able Scottish mathematician of the 18th century.

19. A quadrilateral whose angles are all right angles.

20. An able Swiss mathematician of the 18th century.

21. Bridge scores are often kept on this.

22. Isaac Newton's title.

23. The unit of measure for automobile speeds.

24. A noncircular rotating disc which gives a to-and-fro motion.

25. The abbreviation for the least common denominator.

It is always interesting to look at old books. You may have been fortunate enough to have seen some of the following in old arithmetic books. What could the author have meant by . . . ?

1.   5673̇42̇452

4.   364 | 8

2.   98̇52̇5476

5.
$$4$$
$$\frac{1}{3} \quad \text{by} \quad \frac{3}{4}$$
$$9$$

3.

9 ⤬ 1
6 ⤬ 4

6.   83 4′ 2″ 5‴

7.
    3 feet 8 inches
    5 feet 4 inches
    ——————————
    18   4
     1   2   8
    ——————————
    19 feet 6 $\frac{2}{3}$ inches

8.

5 ——————— 25
8 —————————

1. *When was it? Who was it?*

This is the story of a well-known man born years ago. He has influenced for many generations the thoughts and the minds of men and women in many different lands.

We can tell you that the first and last digits of the year during which he was born add up to the second digit, and that the third digit is one larger than the second digit, and that three times the fourth digit equals two times the third digit.

Can you calculate the year of his birth? Who is this gentleman?

2. *Who caught the bus?*

Juliette and her sister Lucile lived together in that beautiful town of Montreux by Lac Leman in the Swiss Alps. In the springtime one of their favorite walks was to go up to the lovely fields of narcissi growing on the mountain slopes nearby.

On one occasion they came to a long straight stretch of road, and at a certain point on it, they left the road and walked at right angles across a field to a large clump of narcissi. Juliette stopped to pick some of the flowers 40 meters away from the road, while Lucile also collected some flowers another meter farther on. Suddenly they looked up to see a bus going along the road to Montreux. When they had decided to ride home, the bus was 70 meters away from the point where they left the road to walk across the field.

They ran at half the speed the bus traveled to the point where they left the road and missed the bus! There is at least one point on that stretch of road where the bus could have been caught.

Can you calculate where they should have run and if both of the sisters could have caught the bus?

3. *How was this done?*

An Arab when he died left to his three sons seventeen camels, giving to the eldest one four ninths, to the second one third, and to the youngest one sixth of them. The three young men sat in front of their house contemplating how they could fulfill their father's wish without killing any of the animals. They did not find a solution to this problem. Suddenly a dervish came riding along on a camel. They asked him to sit down with them for a moment and told him of their troubles. The dervish pondered for a moment, smiled cunningly, and said, "I know how you can carry out your father's wish without having to kill even one of the animals."

Can you guess what suggestion the dervish made?

4. *Can a sheet of paper have one side only?*

The page on which this is printed has two sides and one edge all the way around. If you tear it out of the book you can easily trace the edge with a pencil. Nevertheless it would be a pity to spoil the book by doing this! If you want to go from one side of the paper to the other, you must go through the paper or over one of the edges.

Can you design a piece of paper that has only one side and also only one edge? If you can do this, then you can paint the whole surface with a brush (if the brush held enough paint) without removing it from the surface or going over an edge.

NUMBER KNOWLEDGE
WITH A DIFFERENCE

*We hope that when you have finished this quiz you have
not "lisp'd in numbers, for the numbers came"! What is the
number of the . . . ?*

1. wonders in the ancient world?

2. Graces in classical mythology?

3. kings in a French deck of cards?

4. beast as given in the New Testament?

5. Muses in Greek mythology?

6. gentlemen of Verona?

7. elements in the teaching of Aristotle?

What is "x" in the following quotations?

8. "Alone and warming her *x* wits
   The white owl in the belfry sits."

9. "*x* days, sire, have elapsed since the fatal moment
   when Your Majesty was forced to quit your
   capital." (The capital is Paris.)

10. "Il faut tourner *x* fois sa langue dans sa bouche
    avant de parler."

On the whole, logarithms have been most useful to those who can add, subtract, and look up tables. They were undoubtedly invented to shorten arithmetical calculations. No tables are needed to answer these questions.

1. Give the meaning of the two Greek words from which the word "logarithm" is derived.

2. What is the logarithm of 16 to the base 4?

3. Were logarithms invented before the time of Newton?

4. Name the two mathematicians who were the cause of an argument between a Swiss and a Scotsman as to who invented logarithms.

5. Who gave us the common logarithms?

6. What nationality was the mathematician who calculated most of the table of common logarithms first published in 1628?

7. Who wrote the *Descriptio?*

8. How can natural logarithms be converted to common logarithms?

9. What useful mathematical instrument is constructed by making use of logarithms?

10. Which is the greatest and which the least of log (2 + 4), (log 2 + log 4), log (6 − 3), and (log 6 − log 3)?

[48]   *Answers on pages 142-144*

Some rather surprising correct results are often found in Charlie's work, which frequently is good only in parts. Here are some examples from Charlie's homework. You have to correct these as quickly as possible. Are they right or wrong?

1. $12^2 = 144$ $\therefore$ $21^2 = 441$

2. $13^2 = 169$ $\therefore$ $31^2 = 961$

3. $\sqrt{5\frac{5}{24}} = 5\sqrt{\frac{5}{24}}$

4. $\sqrt[3]{2\frac{2}{7}} = 2\sqrt[3]{\frac{2}{7}}$

5. The lines joining the mid-points of the sides of a parallelogram form a parallelogram. Therefore the lines joining the mid-points of the sides of any convex quadrilateral also form a parallelogram.

6. $\text{Sin}\,(a + b) \cdot \sin\,(a - b) = (\sin a + \sin b)$
$$(\sin a - \sin b)$$
$$\therefore\ \sin\,(a + b) \cdot \sin\,(a - b) = \sin^2 a - \sin^2 b$$

7. Solve $\dfrac{x - 2}{y - 1} = \dfrac{3}{5}$ and $\dfrac{x - 1}{y} = \dfrac{2}{3}$

$$\therefore\ \frac{x - 1}{y} = \frac{x}{y} - 1 = \frac{2}{3}$$

$$\therefore\ \frac{x}{y} = \frac{2}{3} + 1 = \frac{5}{6}$$

$$\therefore\ x = 5,\ \text{and } y = 6$$

8. How many triangles are there in this figure?
   There are twelve lines.
   Each triangle has three sides.

$$\therefore\ \frac{12}{3} = 4$$

$$\therefore\ 4 \times 4 = 16\ \text{triangles}$$

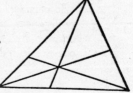

[49]   *Answers on pages 145-146*

CAN YOU ARRANGE THESE?

1. A boy is to be chosen president and a girl vice-president of the senior class of a school. In how many ways is this possible if the class has twelve boys and ten girls?

2. Six boys are to be photographed in a row. How many different arrangements can be made of the order in which they are to sit?

3. The same six boys are to sit around a table for lunch. How many different arrangements can be made of the order in which they are to sit?

4. If the first three letters of a telephone number indicate the name of the exchange, how many such arrangements of three letters is it possible to devise from the twenty-six letters of the alphabet?

5. How many different forecasts must be made of four football games in order to ensure that one forecast is correct?

6. In how many different ways can two dice, one red and one blue, come up when thrown?

7. One of the crews in the Harvard-Yale race has a problem for its captain. Three of the crew are stroke-side oarsmen only and two of them are bow-side oarsmen only. Ignoring weights and personal preferences, in how many ways can the captain arrange his eight men to form the crew? The cox is selected and does not change.

1. A water lily doubles itself in size each day. From the time its first leaf appeared to the time when the surface of the pond was completely covered took forty days. How long did it take for the pond to be half covered?

2. A quart bottle had all its dimensions doubled. What is the volume of the new bottle?

3. From Philadelphia to Atlantic City is 60 miles. Two trains leave at 10:00 A.M., one train from Philadelphia at 40 miles an hour and the other from Atlantic City at 50 miles an hour. When they meet, are they nearer to Philadelphia or to Atlantic City?

4. Spot the wrong number in these series of numbers:
   (a) 1, 2, 4, 8, 15, . . .
   (b) 1, 7, 27, 64, 125, . . .
   (c) 10, 15, 21, 25, 30, . . .

5. What is the missing number in these series:
   (a) 81, 27, —, 3, 1, . . .
   (b) 1, 4, 9, —, 25, . . .
   (c) 2, 6, 12, —, 30, . . .

6. If a half dollar is placed on the table, how many half dollars can be placed around it touching it and each other?

7. Write down the Roman numerals from "one" to "six" as seen on a clock face.

QUIZ NO. **39** *VACATIONS ABROAD—
BUT WHERE?*

*"The picture of contentment" shows a family gathered round
a warm log fire on a cold winter's evening. What are they
doing? Planning next year's summer vacation! Letters are
written. In some of the replies the following were included,
and you have to name the country they wrote to.*

1. We can give you full board at eight-five pesetas per
   day.

2. The cheapest suitable fare we can offer is at forty
   bahts a day.

3. You will be comfortable and well-fed at seventy-
   four schillings a day.

4. Our rate is eighty-four drachmae per day.

5. One thousand yen per day is our cheapest rate.

6. About sixty escudos is the normal cost.

7. The inclusive cost is seventeen hundred and fifty
   lire a day.

8. Fifty-four soles per day is cheap for good fare.

9. You will find the usual cost in a good hotel to be
   fifteen kyats a day.

10. Twenty levas a day is the price we quote.

The answer to each question below will give you a letter. If these are written down they will form an anagram of the name of a well-known baseball team.

1. What is the abbreviation for "pence" in England?

2. $1 \cdot 2 \cdot 3 \cdot 4 \ldots (n-2) \cdot (n-1) \cdot (n) = ?$

3. What represents the first term in the formulae for an arithmetical progression and a geometrical progression?

4. For a rectangle $1 \times b = ?$

5. For a circle $\pi d = ?$

6. What is the symbol in trigonometry for one half of the sum of the lengths of the sides of a plane triangle?

7. What is the square root of -1?

8. In the series 9, 27, 81, etc. $? = 3$?

9. What is the abbreviation for 1,000 cubic centimeters?

 Answers on page 152

```
                    1
                 1     1
              1     2     1
           1     3     3     1                   4    9    2
        1     4     6     4     1
     1     5    10    10    5     1              3    5    7
  1     —     —     —     —     —     1
1     7    21    35    35    21    7     1       8    1    6
1     8    28    56    70    56    28    8     1
1  —   —    —    —    —    —    —    —    1
```

1. Write down the seventh line of figures in the arithmetical triangle.

2. What are the missing numbers in the last line of the arithmetical triangle?

3. Where in the arithmetical triangle do the coefficients of the terms of $(x + a)^2$ and $(x + a)^3$ appear?

4. Use the triangle to work out the coefficients of $(x + 2)^4$.

5. Who is the mathematician associated with this triangle?

6. Find the sum of the numbers in each column, each row, and each diagonal of the square printed above. What name is given to a square built in this way?

[54]

7. Complete a number square built in the same way as the one printed above, given:

16   2   12

6   —   —

8   —   —

8. Construct a number square of four rows and four columns such that the sum of each column, row, and diagonal is the same, and given that the top row is 1, 15, 14, and 4, and the left-hand column is 1, 12, 8, and 13.

1. The first five terms of the series 10, 20, 30, 40, 50 add up to 150. What five terms of another series, without fractions, add up to 153?

2. Find three vulgar fractions of the same value using all the digits 1 to 9 once only. Here is one solution of the problem:

$$\frac{3}{6} = \frac{7}{14} = \frac{29}{58}$$

3. A boy selling fruit has only three weights, but with them he can weigh any whole number of pounds from 1 pound to 13 pounds inclusive. What weights has he?

4. Can you, by adopting a mathematical process, such as $+$, $-$, $\times$, $\div$, $\sqrt{}$, etc., use all and only the digits 9, 9, 9 to make (a) 1, (b) 4, (c) 6?

5. From where on the surface of the earth can you travel 100 miles due south, then 100 miles due west, and finally 100 miles due north to arrive again at your starting point?

6. A train traveling at 60 miles an hour takes three seconds to enter a tunnel and a further thirty seconds to pass completely through it. What is the length of the (a) train, (b) tunnel?

QUIZ NO. 43 ## WHAT CAN THEY REFER TO?

In this quiz you can test your general knowledge with special reference to mathematics. This is not intended to test the mathematical genius, and some of us may be pleased that no calculations are involved. All the answers are very brief, so it is useless to attempt to cover up ignorance with a host of words! It is tantalizing to have the answer on the tip of your tongue and yet to be unable to give it. So go to it and see if your general knowledge is up to the mark. What do the following refer to?

1. "Noe two thynges can be moare equalle."

2. Ludolph's number.

3. Cossic art.

4. The sieve of Eratosthenes.

5. The golden section.

6. The ambiguous case.

7. A gnomon or style.

8. Casting out the nines.

9. The curve of quickest descent.

10. A soraban.

Here you are faced with a succession of terms or quantities which, after the first term or quantity, are formed according to a common law. This sounds very complicated, but one grain of common sense plus two grains of confidence is all that is necessary to have some fun with the following series.

1. My reciprocals are in arithmetical progression, and I hope I am of some interest in the theory of sound. What is my name?

2. The ratios of successive terms of this series are connected with plant growth. The leaves of a head of lettuce and the layers of an onion grow like this. What is my name?

3. What is the sum of the first twenty terms of this series?

$$1 + 3x + 5x^2 + 7x^3 + \cdots$$

4. What is the eighth term and also the sum of the first eight terms of this series?

$$5 \cdot 7 \cdot 9 + 7 \cdot 9 \cdot 11 + 9 \cdot 11 \cdot 13 + 11 \cdot 13 \cdot 15 + \cdots$$

5. Is the logarithmic series,

$$\log_e (1 + x) = x - \frac{x^2}{2} + \frac{x^3}{3} - \frac{x^4}{4} + \cdots$$

useful for working out logarithms to the base *e*?

6. What is the name of this series?

$$x - \frac{x^3}{3!} + \frac{x^5}{5!} - \frac{x^7}{7!} + \cdots$$

7. What is the name of this series?

$$1 + x + \frac{x^2}{2!} + \frac{x^3}{3!} + \cdots + \cdots$$

GIVE THE CREDIT
WHERE IT IS DUE

In this quiz the term "Father" is used and its meaning includes not only the mathematician who conceived the first ideas on a particular subject or invented a piece of apparatus, but also the author of a mathematical book or treatise. Thus Newton is called the Father of the Principia. Who is the Father of . . . ?

1. geometry?

2. the great work of thirteen "books" called the *Elements?*

3. the requirement that geometrical constructions be confined to a ruler and a compass?

4. our notation of the calculus?

5. modern number theory and the electromagnetic telegraph?

6. *The Grounde of Artes?*

7. some special "numbers" published in his *Ars Conjectandi* in 1713?

8. the complete solution of the problem of a vibrating string in sound, and the *Mécanique analytique?*

9. the world map made by projecting a spherical surface on a plane?

10. the papyrus with the title "Directions for knowing all dark things"?

ACROSS

1. We have to "tot" these up.
2. Set for the unwary in exams.
3. "The pointers" always point to one of these.
4. Reversed abbreviation for one chord of a parabola.
5. The beginning and end of 17 down.
6. Roman figures for 55.
7. Did it really impress Newton?
8. This looks perfect on your bank statement.
9. The initials of a mathematician who put forward the laws governing the forces of gravitation.
10. The area of a rectangle.
11. 2.54 of these equal 1 inch.
12. The unit, 33,000 footpounds per minute.
13. The obvious use for a scale.

DOWN

3. The discoverer of the laws of refraction of light.
14. The monetary unit of Yugoslavia.
15. A frequent incorrect abbreviation for inches.
16. Given when money is borrowed.
17. Gives orientation on a plan.
18. He wrote the oldest extant Greek mathematical treatises.
19. Graphs resembling histograms.
20. A useful table for surveying.
21. Used by the Egyptians to mark out their fields.
22. A positive number.
23. The metric system of units.
24. Indicates a negative characteristic in logarithms.
25. A trigonometrical ratio which is written as if it were evil.

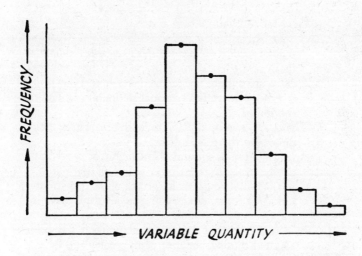

What is the name of . . .

1. this special column graph?

2. the shape formed by joining the mid-points of the tops of the columns?

3. the frequency curve shaped like a cocked hat?

4. the arithmetical average of the values of a variable quantity?

5. the most frequently observed value of a variable quantity?

6. that which most satisfactorily indicates the spread of the observed values of a variable quantity?

7. the sample chosen such that every sample has an equal chance of being picked?

[61] *Answers on pages 167-168*

1. How far can you go into a forest?
2. A man drives along a main highway on which a regular service of buses is in operation. He notices that every three minutes he meets a bus and that every six minutes a bus overtakes him. How often does a bus leave the terminal station at one end of the route?
3. There are twelve dollars in a dozen. How many dimes are there in a dozen?
4. An airplane flies around the equator at a constant height of 200 feet. If the radius of the earth is 4,000 miles how much farther than the circumference of the earth will the airplane have to travel?
5. In a small town of 50,000 inhabitants, it has been counted that 42 per cent of the males and 28 per cent of the females married people from their own town. Assuming these numbers have remained fairly constant over the years, how many males are there in the town?
6. You are standing at the center of a circle of radius 9 feet. You begin to hop in a straight line to the circumference. Your first hop is $4\frac{1}{2}$ feet, your second $2\frac{1}{4}$ feet, and you continue to hop each time half the length of your previous hop. How many hops will you make before you get out of the circle?
7. Three students have two boxes of candy which they want to share equally among themselves. Neither the number of pieces in the first box nor the number in the second is divisible by three. Yet one of the students noticed that there were seven more pieces in the second box than in the first and then he said, "We can share this candy equally between us." Was he correct?

CIRCLES, CIRCLES, AND
MORE CIRCLES

Given the following clues, can you name the circle which is implied?

1. It seems to be "a manager," but the two tangents from any point on it to an ellipse are at right angles.

2. It seems as if this circle could be helpful to an ellipse.

3. Is a circle very much tied up with the feet of the altitudes and the mid-points of the sides of a triangle. What size shoe did Clementine wear?

4. Two circles which cut "right" across each other.

5. The circle which touches all the sides of a polygon.

6. King Alfred did not really name this circle!

7. The circle which seems to be suffering from "spring fever."

8. A triangle is greedy enough to have more than one of these circles.

9. A circle which passes through the vertices of a triangle.

There are some problems which anyone with an elementary knowledge of calculus can solve with the utmost ease. After a short time a beginner can tackle interesting and practical problems. The feeling of accomplishment, and even fun, which the subject brings when rightly used is also increased by the beauty of its methods. Below are four problems which can be solved by the aid of this admirable instrument.

1. A hiker on the moors is 2 miles from the nearest point, P, on a straight road. 8 miles from P along the road is an inn. The hiker can walk at 3 miles per hour over the grassy moors and at 5 miles per hour along the good road. At what distance from P must he aim to strike the road in order to get to the inn as quickly as possible?

2. Dan Dare the space-ship pilot wears a space hat in the shape of a paraboloid of revolution. The diameter of the circular base is 8 inches and the height of the hat is 12 inches. What volume of heavy water will it hold?

3. Equal squares are cut out at each of the corners of a rectangular sheet of tinfoil whose dimensions are 32 inches by 20 inches. Find the maximum volume of a wooden box which can be lined by suitably bending the tinfoil to cover the base and the sides of the box.

4. A pleasure steamer 150 feet long has changed its direction through 30 degrees while moving through a distance equal to twice its own length. What is the radius of the circle in which it moved?

ANSWERS

ANSWERS TO QUIZ NO. 1

1. Yards in 1 mile.

2. Pounds in 1 ton.

3. Square yards in 1 acre.

4. Acres in 1 square mile.

5. The square root of 3.

6. Centimeters in an inch.

7. π—the ratio of the circumference of a circle to its diameter.

8. Days in a leap year.

9. The logarithm of 2 to the base 10.

10. The year in which Christopher Columbus found land (in the Bahamas) by sailing west from Spain.

11. The logarithm of 3 to the base 10.

12. The logarithm of π to the base 10.

13. 1.6 kilometer = 1 mile, and 0.6214 mile = 1 kilometer.

14. The square root of 2.

15. Cubic inches in 1 cubic foot.

16. Ratio of the sides of a right-angled triangle.

17. Feet in a nautical mile.

18. 62½ pounds is the weight of 1 cubic foot of water.

19. Degrees in 1 right angle.

20. 88 feet per second is the same as 60 miles per hour.

1. 2228
    1261
    ———
    3489
    ———

    113
    624
    147
    526
    ———
    1410
    ———

2. 4472
    4351
    ———
    121
    ———

    6235
    5828
    ———
    407
    ———

3.   47
      53
    ——
    141
    235
    ——
    2491
    ——

    5467
     898
    ———
    43736
    49203
    43736
    ———
    4909366
    ———

4. $5x - 5 = 4x - 3$
    $\therefore x = 2$

    $x^2 - 4x = 21$
    $\therefore x = 7 \text{ or } -3$

ANSWERS TO QUIZ NO. 3

1. =

Both the Greeks and the Arabs used a letter for "equals." In the Middle Ages the full word was employed, and in the seventeenth century Descartes used the variation symbol to denote equality. Robert Recorde in his algebra book introduced the two familiar short lines in 1557.

2. ×

The multiplication symbol was probably first introduced by William Oughtred. It had previously been used for other purposes in mathematics, but one can see why it was not adopted in algebra, because of its resemblance to the letter "x." By the latter half of the nineteenth century it was used in elementary arithmetic.

3. $\sqrt{}$

Ancient writers usually wrote the word "root," and this practice was followed by the use of the letter "r." The first known use of the present symbol was made by a German named Rudolff in 1526, and it is said that he invented it because the symbol resembles the letter "r." It was not universally adopted until the seventeenth century.

4. $\propto$

This sign has been used for the symbol of equality, as also has the mirror-image of it. In science the words "varies as" are often used and the mathematical symbol "$\propto$" means precisely this. If $a \propto b$, then it follows that $a = k \cdot b$, where k is a constant.

5. >

Oughtred had suggested symbols for "greater than" and "less than," and hence when the present symbols were proposed by his contemporary Thomas Harriot, an English mathematician, they were not readily accepted. After they were used in print in 1631, they slowly gained favor. However, Oughtred's symbols were still in use in the eighteenth century.

6. ∞
This symbol was used for "ten thousand" by the Greeks. The symbol representing a large number was called infinity. In 1665, John Wallis took the symbol for "ten thousand" to be the symbol for a large number.

7. ::
William Oughtred is also responsible for this proportion symbol, which appears in his *Clavis Mathematicae*, an algebra book published in 1631. Curiously enough, Oughtred used a dot to denote a ratio, the symbol now used for a decimal point.

8. $_nC_r$
The interest in selections goes back to the time of the early Chinese and Hindu mathematicians, but it was not until the late seventeenth century that the word "combination" was used in its present sense. Strictly speaking, $_nC_r$ is an abbreviation and not a symbol.

9. π
Although this symbol was first used in 1706 to represent the ratio $\dfrac{\text{circumference}}{\text{diameter}}$ for any circle, it was not widely adopted until late in the eighteenth century. The Swiss mathematician Euler was attracted by it and did much to increase its popularity. Previously, π was used to denote the periphery of a circle.

10. a^3
It was René Descartes who in 1637 wrote aa or a^2 to mean multiplying a by itself and a^3 to mean the product of a^2 and another a. John Wallis and Isaac Newton explained the ideas of negative and fractional indices, and thus extended the invention of Descartes.

ANSWERS TO QUIZ NO. 4

1. FIVE GIRLS

Five girls pack five boxes in five minutes,
Five girls pack one box in one minute (working on the same box!),
Five girls pack fifty boxes in fifty minutes.

2. 47 SECONDS

The time taken will be 47 seconds, because the 47th cut produces the last two squares.

3. 44 INCHES

Steel bars are often used as rollers in this way. The safe moves forward twice the length of the circumference of one of the steel bars. This distance is therefore $\frac{2 \cdot 22 \cdot 7}{7}$ inches, which is 44 inches. With three or any number of rollers under the safe it will still move forward 44 inches. The best way to see this is to consider this problem in two parts:

(a) the motion forward caused by one revolution of the rollers if they were rolling off the ground,

(b) the motion forward of the centers of the rollers because they themselves roll forward on the ground.

In both cases the motion amounts to 22 inches, so that the total movement of the safe mounted on the rolling rollers is 44 inches.

4. 20,000

Did you get this right? It really does not matter what percentage of the population is one-legged! All the one-legged people will require only one shoe in any case. Of the remainder, half will wear no shoes and the other half will carry two shoes on their two feet. This works out at one shoe per person for the "others." We shall therefore need for the whole population on the average one shoe per person.

5. $99\dfrac{99}{99}$

This is one of the old trick problems. If recognized signs were introduced, you could arrange six "nines" to give 100 as follows:

$$[(9 \times 9) + 9] + 9\frac{9}{9}$$

You could also arrange four "nines" to give 100 as follows:

$$99\frac{9}{9}$$

6. ALL THE NUMBERS 0 TO 9 APPEAR

This is interesting, but it is by no means the only example of composing 100 from all the numbers 0 to 9 taken once only. Examine these solutions:

(a) $0 + 1 + 2 + 3 + 4 + 5 + 6 + 7 + (8 \times 9)$
(b) $78\frac{3}{6} + 21\frac{45}{90}$
(c) $89 + 6\frac{1}{2} + 4\frac{35}{70}$
(d) $90 + 8\frac{3}{6} + 1\frac{27}{54}$
(e) $1 + 2\frac{35}{70} + 96\frac{4}{8}$
(f) $97\frac{30}{45} + 2\frac{6}{18}$
(g) $97\frac{43}{86} + 2\frac{5}{10}$

Can you design still more solutions?

7. 128 INCHES

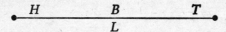

Let H represent the head, B the body, T the tail, and L the total length of the fish.

Looking at the problem we are given the following three facts:

$$T = H + \tfrac{1}{4}B$$
$$B = \tfrac{3}{4}L$$
$$H = 4 \text{ inches}$$

It is also true that $L = H + B + T$

In this equation keep L and substitute for everything else.

$$\therefore \ L = 4 \text{ inches} + \tfrac{3}{4}L + (H + \tfrac{1}{4}B)$$
$$\therefore \ L = 4 \text{ inches} + \tfrac{3}{4}L + 4 \text{ inches} + \tfrac{3}{16}L$$
$$\therefore \ L = 8 \text{ inches} + \tfrac{15}{16}L$$
$$\therefore \ \tfrac{1}{16}L = 8 \text{ inches}$$
$$\therefore \ L = 8 \times 16 \text{ inches}$$
$$\therefore \ L = 128 \text{ inches}$$

Thus we see that the fish was 128 inches long.

1. THE ABACUS

The merchants and traders of ancient days in Egypt and Mesopotamia used to set out pebbles in grooves of sand to calculate and add up accounts. There would be a "units" groove, a groove for "tens," and one for "hundreds." Such was a simple abacus, and the word is derived from a Greek word meaning "tablet." In Roman times, a calculating frame was made in which pebbles slid on wires and this was also called an abacus. The size of the abacus determined the size of the numbers which could be dealt with. The Roman numerals made simple addition, subtraction, and multiplication very complicated. Calculations were done by slaves using an abacus. It is interesting to note that the Roman word for "pebble" was "calculus" and here we have the derivation of our word "calculate."

2. 1776

Letters were used by the Romans to represent various numbers, and they seem to follow the pattern set by the Greeks.

1, 5, 10, 50, 100, 500, 1000
I, V, X, L, C, D, M

The Roman numerals on the monument therefore read as:

M = 1000, D = 500, C = 100, C = 100, L = 50, X = 10, X = 10, V = 5, I = 1

These added together make 1776, and every American knows this date!

3. MDCCLXXXIX

M = 1000, D = 500, C = 100, C = 100, L = 50, X = 10, X = 10, X = 10, and IX = 9. Add these all together and the result is 1789.

4. CLXVI

This number is 166.

5. CXLIV
This number is 144. XL is ten before fifty, which is forty. Similarly, IV is one before five, which is four.

6. CD

C	X	VI
	X	III
		VI
CC	LX	V

CD

The answer of the addition is four C's, which is four hundred. It was the custom not to write four similar numerals consecutively. Hence, instead of writing four "hundreds" (CCCC), the Romans wrote one hundred less than five hundred (CD). Placing the C before the D meant C less than D, and placing it after the D, as in DC, meant C more than D. So that CD is four hundred and DC is six hundred.

7. 4 and 9
The Roman numeral for four is IV. Therefore, when the left-hand integer is removed there remains the integer V. V is the Roman numeral for five. Hence the four changes to five. Similarly, the Roman numeral for nine is IX, and when the left-hand integer I is removed there remains the Roman numeral X, which is ten.

8. SURVEYOR
The "groma" was an important surveying instrument used by the Roman surveyors or agrimensores. As far as mathematics was concerned, the Romans were practical and no more. To them mathematics was a tool that helped them to construct and to measure. The agrimensor was a land- or field-measurer, and in this work he used the groma. It was frequently carved on the tombstones of Roman surveyors.

ANSWERS TO QUIZ NO. 6

1. 18 *DAYS*
There are three possible types of day:
    (a) Rain in the morning and fine in the afternoon
    (b) Fine in the morning and rain in the afternoon
    (c) Fine in the morning and fine in the afternoon

Let the number of such days in each category be a, b, and c.
$\therefore$ number of days on which rain falls $= a + b = 13$
$\therefore$ number of days having fine mornings $= b + c = 11$
$\therefore$ number of days having fine afternoons $= a + c = 12$

From these equations, we derive that $a = 7$, $b = 6$, and $c = 5$.
$\therefore$ number of days on vacation is $7 + 6 + 5 = 18$.

2. $5\frac{5}{11}$ *MINUTES PAST* 7
At 7 o'clock the minute hand is 35 divisions behind the hour hand.
To be opposite one another the minute hand must gain 5 divisions on the hour hand.
But the minute hand gains 55 divisions in 60 true minutes.
$\therefore$ the minute hand gains 5 divisions in $5\frac{5}{11}$ true minutes.
All problems concerning the positions of the hands on clock faces will be solved readily if you draw a sketch and remember that the accurate position is "something" and "something"-elevenths.

3. 113
1,442,897 is a seven-figure number and thus the cube root must lie between 110 and 120. As the last figure is a 7, the cube root must end in 3.

4. FIVE CENTS

Let your algebra help you!

$$C + B = 25$$
$$C = 15 + B$$
$$\therefore B = 5.$$

5. 9, 40, *and* 41 FEET

The procedure is as follows: square the length of the perpendicular, subtract 1, divide by 2. The result is the length of the base. Add 1 and then that is the length of the hypotenuse. This applies when the perpendicular is any odd number and these combinations of numbers are sometimes called Pythagorean series. Other combinations are 3-4-5, 5-12-13, 7-24-25, 11-60-61, 13-84-85, and so on indefinitely. The method is derived from the theorem of Pythagoras concerning a right-angled triangle.

$$H^2 = B^2 + P^2$$
$$\therefore (B + 1)^2 = B^2 + P^2$$
$$\therefore 2B + 1 = P^2$$
$$\therefore B = \frac{P^2 - 1}{2}$$

6. 15 FEET

Let the tree grow x feet each year.
At the end of the sixth year, the height of the tree

$$= (3 + 6x) \text{ feet}$$
$$\text{The growth } x = \tfrac{1}{6}(3 + 6x)$$
$$\therefore x = \tfrac{1}{3} + \tfrac{2}{3}x$$
$$\therefore x = 1$$

At the end of the twelfth year, the height
of the tree $= (3 + 12x)$ feet
$$= 15 \text{ feet.}$$

7. YES

Do you know the tests of divisibility? Numbers will divide exactly

by 2 if they end with an even digit
" 3 if the sum of the digits is divisible by 3
" 4 if the last two digits are divisible by 4
" 5 if the last digit is 0 or 5
" 6 if divisible by both 2 and 3
" 8 if the last three digits are divisible by 8

What about the tests of divisibility for 7, 9, 11, and 12?

8. GRADE A: GRADE B = 3 : 1

Let A parts of Grade A be mixed with B parts of Grade B cement, then the equation is:

$$36A + 24B = 33 (A + B)$$
$$\therefore 3A = 9B$$
$$\therefore \frac{A}{B} = \frac{3}{1}$$

ACROSS

1. 1,728

3. 33

5. 26. Arithmetical mean $= \dfrac{50 + 2}{2}$

6. 981

7. 5,280

10. 1,890. Simple interest $= \dfrac{P \times r \times T}{100} = \dfrac{18,900 \times 5 \times 4}{100 \times 2}$

12. 144

One girl must sit down in any place and then the other boys and girls can be arranged around her at the table.

The seat next but one to her on her left can be occupied by any one of three girls . . . that means in 3 ways.

The seat opposite to her can then be occupied by either one of the two remaining girls . . . that means in 2 ways.

The seat next but one to her on her right can only be occupied by the last remaining girl . . . that is in 1 way.

∴ the girls can be seated in $3 \times 2 \times 1$ or 3! or 6 ways.

Similarly the boys can be seated in $4 \times 3 \times 2 \times 1$ or 4! or 24 ways.

∴ the girls and boys can be seated in 6×24 or 144 ways.

14. 45. The regular octagon has 8 equal sides

∴ the angle at the center is $\dfrac{360}{8} = 45°$

16. 25. Diagonal^2 = Sum of squares on the other two sides
$$= (3 \times 5)^2 + (4 \times 5)^2$$
$$= (5 \times 5)^2$$

17. 1,458. This series has a common ratio of 3

∴ the sixth term is $6 \times 3^5 = 6 \times 243$

DOWN

1. 12

2. 765. (Note: 88 ft. per sec. = 60 m.p.h.)

3. 38. (Note: Opposite angles of a cyclic quadrilateral = 180°)

4. 3,142

6. 9,009. A number is divisible by 11 if the sum of the digits in the even places equals the sum of the digits in the odd places.

8. 2,814. A quick rule to find this answer is to add π times the width to the internal circumference and then multiply by the width. Why is this so?
Area = $[(3\frac{1}{7} \times 10 + 250)]$ 10

9. 1812. To find whether a year is a leap year, divide the number formed by the last two digits by 4, except when they are both zeros. If they are both zeros, divide the first two digits by 4. In each case if there is no remainder then the year is a leap year. 1804, 1808, and 1812 are leap years, but 1800 is not.

11. 045. Directions in air navigation are given as angles of three figures in a clockwise direction from north.

13. 45°

15. 58. The sides of this triangle are in the ratio of 1, $\sqrt{3}$, and 2.

[80]

ANSWERS TO QUIZ NO. 8

1. EQUILATERAL

As the name suggests, all the sides are equal. Equilateral triangles have not only equal sides but have equal angles too and can thus be called equiangular triangles. The trillium has three axes of symmetry suggesting an equilateral triangle. Many plants have geometrical shapes in their roots, stems, leaves, and flowers. What shape is found in the cross-section of the stem of the sedge family?

2. ISOSCELES

The word means "equal legs." Any triangle which has two of its three sides equal is called an isosceles triangle. If we look around us we can often find geometrical figures in nature. In the construction of houses, ships, and aircraft, the isosceles triangle is often encountered.

3. SCALENE

This is a triangle which is "uneven" because it has all its sides unequal. The word "scalene" has nothing to do with drawing to scale, but is derived from the Greek word *"skalenos,"* which means "uneven" or "unequal."

4. PEDAL

This triangle is sometimes called the orthocentric triangle. If *AD*, *BE*, and *CF* are the perpendiculars dropped from the vertices of the triangle *ABC* to the opposite sides, then the triangle *DEF* is the pedal triangle. The three perpendiculars pass through a common point called the orthocenter.

5. RIGHT-ANGLED

The unique property of the right-angled triangle *ABC* is that if angle *A* is the right angle then $a^2 = b^2 + c^2$. The geometrical proof of this property is associated with the Greek mathematician Pythagoras, who lived in the sixth century B.C. Triangles whose sides are in the ratios 3-4-5 or 5-12-13 are right-angled.

6. SIMILAR

Two triangles are said to be similar if they have their angles equal to one another and have their sides, taken in order, about the corresponding equal angles proportional. The areas of similar triangles are proportional to the squares of their corresponding sides.

7. HARMONIC

ABCD is the quadrilateral and OXY is the harmonic triangle. It is formed by joining the intersection points of the sides and the intersection point of the diagonals.

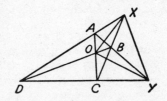

8. SPHERICAL

A spherical triangle consists of a portion of a sphere bounded by three arcs of great circles. Obviously the sides do not consist of straight lines. The sum of the three angles of such a triangle lies between 180° and 540°. Much of the early work with spherical triangles was done by Menelaus about A.D. 100.

9. ISOSCELES

The particular proposition proved that the angles at the base of an isosceles triangle are equal. The title "pons asinorum" means the "bridge of asses." It is said that in the Middle Ages the "donkey" could not pass over this bridge to continue his study of Euclidean geometry, but the name may be due to the fact that the figure in Euclid resembles a simple truss bridge.

10. ASTRONOMICAL

This is a spherical triangle on a celestial sphere which has for its vertices the nearest celestial pole, the zenith, and the star under consideration.

ANSWERS TO QUIZ NO. 9

1. CIRCLE

The use of the phrase "circle of influence" seems to go back as far as the late seventeenth century. Later it seems that it was considered that either a person or an object exerted its influence on all planes and hence the phrase "sphere of influence." We should also note the use of the word "circle" with reference to a group of persons surrounding a center of interest. What is the family circle?

2. SQUARE

To the golfer who has been trailing against an opponent for nearly a complete round, the words "all square" will come as sweet music. The word "square" was used in the sixteenth century to mean concur, correspond, or agree with. In 1887 golfers began using the term with reference to equal scores.

3. ZERO

It is from the zero mark on a graduated scale that reckoning begins. Hence we can see how the military term "zero hour" came into being. It means the precise time at which an operation or attack is planned to begin. It became a term in general use in the first World War and has filtered through into our more sensible and civilian life with the same meaning. All students face a zero hour with their examinations!

4. DIVIDE

We, in America, all know the particular use of this word to mean a ridge or line of high ground forming the division between two river valleys or systems—a watershed. "The Great Divide."

5. PROPORTION

The phrase intended here is "out of all proportion." How often do we consider a punishment to be out of all proportion or excessive for the particular offense committed? Proportion in the mathematical sense is an exact equality of ratios.

6. QUOTIENT

In mathematics the quotient is obtained by dividing one quantity by another. In order to have a unit for intelligence, it has been found that the most useful is that obtained by dividing the mental age by the chronological age and multiplying the result by 100. This quantity is called the I.Q. or "intelligence quotient." We hope yours is over 100!

7. PENTAGON

It is on record that in 1571 a fort with five bastions was called a pentagon. Thus the use of a "Pentagon Building" for a military establishment has a precedent in history. In mathematics a plane rectilinear figure having five sides and having five angles is called a pentagon. In the regular pentagon the five sides are equal.

8. EXPONENT

In algebra the word "exponent" is used to describe the symbol denoting a power and it is sometimes called an index. In everyday use "exponent" is used of one who expounds or interprets.

9. NEGATIVE

In the seventeenth century "negative" was first employed to denote quantities to be subtracted from other quantities, but the sign we use for this operation did not become popular until some years later. The term is in general use, of course, with reference to rejection and the veto.

10. SUMS

A judge "sums" up the evidence or recapitulates briefly the facts proven during the trial. A "sum" in mathematics is the result obtained by addition or, more loosely, refers to a problem in arithmetic.

1. *ARCHIMEDES*

He was one of the earliest mathematicians and lived in the third century B.C. He attended lectures at Alexandria University and returned to Sicily, his birthplace, where he spent the rest of his life. He possessed great ability and is well known because of his inventions. One of his famous achievements is the Archimedean screw. Some of his writings still exist and these include works on the circle, the spiral, the sphere, the cylinder, and arithmetic.

2. *DESCARTES*

He was a contemporary of Galileo and was born near Tours in France in 1596. As a mathematician he is best known for his contributions to analytical geometry. Everyone who has drawn a graph has used Cartesian co-ordinates, named after Descartes. He was the first person to announce this useful discovery. Descartes published works on algebra and astronomy as well as the first treatise on co-ordinate geometry, which was called *La Géométrie*.

3. *MENELAUS*

Menelaus of Alexandria lived in the first century A.D. He is well known for his extant work on spherical trigonometry. His writings on plane trigonometry are unfortunately lost. In more advanced geometry we generally learn after Ceva's theorem another theorem attributed to Menelaus. This deals with a transversal meeting the sides of a triangle internally or externally.

Almost everybody has heard of Halley's comet; it is interesting to note that it is the same Halley who edited the works of Menelaus.

4. *CARDANO*

This Italian name is often spelled as Cardan in English. He lived in the sixteenth century and won fame for his published works on arithmetic and algebra. Another mathematician, Tantaglia, had discovered a solution of cubic equations, which he passed on in confidence to Cardano. The latter, however, published it in his treatise on algebra.

5. NAPIER

John Napier was a Scottish mathematician who lived in the latter half of the sixteenth century and the early years of the seventeenth century. His discovery of logarithms would alone give him fame. He invented a means of making calculations by means of "bones" or "rods." To Napier is also frequently attributed the honor of being the first to write decimals with a decimal point. This practice is something we readily accept without giving a thought to the discoverer.

6. WALLIS

John Wallis was another seventeenth-century mathematician, who became Savilian professor of geometry at Oxford. He was a profuse writer and his treatise on algebra is most comprehensive. Wallis was one of the founders of the Royal Society in 1645.

7. OUGHTRED

A seventeenth-century English mathematician, who published a good textbook on arithmetic in which he introduced new symbols. To Oughtred is accredited the first use of the abbreviations "sin," "cos," and "tan." The invention of the slide rule is also due to him. He was a good teacher and corresponded with most of the leading mathematicians of his time. It is reported that his death was caused by the excitement of hearing of the restoration of Charles II to the throne.

8. LAPLACE

He was born in Normandy in 1749, and he is sometimes known as the "Newton of France" for he was a great theoretical astronomer. He had the gift of being able to apply mathematical methods to astronomy, molecular physics, electricity, and magnetism. His complete works have been published in fourteen volumes. He discovered the invariability of the major axes of the planetary orbits, and explained the movements of the planets Jupiter and Saturn. In this and other ways he solved many of the problems of the solar system. He introduced the famous nebula theory, in which he maintained that the solar system was the result of a contracting nebula. His work called *Traité de Mécanique Céleste* deals with the motion of the solar system. He discovered a famous differential equation, which has since been named after him.

9. EULER

He was an eighteenth-century Swiss mathematician who was closely connected with the Bernoullis. Euler was a prolific writer on all sorts of mathematical subjects, and even after he became blind he continued to work. He was responsible for the symbol *"e"* which is the base of natural logarithms. Euler used this symbol in 1731. He made the present trigonometrical abbreviations and the use of π more general.

10. DEMOIVRE

His name is more correctly written de Moivre. He was an eighteenth-century French mathematician who lived most of his life in London. The *Principia* made him interested in mathematics and among his contributions we have Demoivre's theorem in trigonometry. He was appointed by the Royal Society to the commission concerning the controversy as to whether Newton or Leibnitz discovered calculus.

ANSWERS TO QUIZ NO. 11

1. ONE INCH

Surely the bookworm has only to go from A to B as in the figure! This distance is the thickness of Volume II—one inch.

2. THE NORTH POLE

One cannot imagine that there is a cubical igloo at the North Pole! But if there were and if it had windows they would all have a southern outlook.

3. 630 LITERS

Let x be the volume of the second barrel. Your equation will be:

$$\frac{5}{6} \times 336 = \frac{4}{9} \times x$$

$$\therefore x = \frac{5 \times 336 \times 9}{6 \times 1 \times 4}$$

4. ELLIPSE, PARABOLA, HYPERBOLA

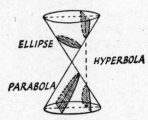

These three curves are often referred to as "conic sections." The sections are shown in the figure—the hyperbola is parallel to the axis, the parabola is parallel to the slant height, and the ellipse is oblique. The base is, of course, a circle.

5. NINE

A wins three games and thus gains three pennies. *B* has to win back these three pennies which takes another three games, and finally *B* wins three more games to win the total sum of three pennies.

6. 4

If $x = $ the missing digit, then the solution is found from this equation:

$$900 + 50 + x - (100x + 50 + 9) = 100x + 90 + 5$$

Thus the number is 954, and $954 - 459 = 495$.

1. *d and* πr^2

The circumference of a circle is π multiplied by *d*, where *d* is the diameter of the circle. It is better to use πd rather than $2\pi r$ because π is the ratio of the circumference to the diameter of the circle. The area of a circle is πr^2.

2. *ARCUS and CHORDA*

The Latin word for a "bow" is "arcus," and for a "string" is "chorda." From these two Latin words we at once recognize the familiar English words "arc" and "chord." These words are first encountered in the geometry of the circle, but the terms are used with reference to all mathematical curves.

3. *FINDING A SQUARE WITH THE SAME AREA
AS A GIVEN CIRCLE*

This was one of the problems which confronted the Greek mathematicians. The difficulty lies in the fact that a ruler and compass only could be used. Attempts to solve the problem go back as far as 460 B.C. . . . if only we could draw a straight line equal to the circumference of a circle!

4. π

This ratio (*see* answer 1 above) is very important and has been investigated since earliest mathematical times, but it was not until 1706 that an English writer, William Jones, definitely used π to mean the same as it does today.

5. *16.13 INCHES*

In 60 minutes the tip of the minute hand of the clock makes one complete revolution tracing out the circumference of a circle whose radius is 7 inches.

In 60 minutes the tip moves through

$$\pi d \text{ inches} \quad \text{or} \quad \frac{22}{7} \times 2 \times 7 \text{ inches}$$

In 22 minutes the tip moves through

$$\frac{22}{7} \times 2 \times 7 \times \frac{22}{60} \text{ inches} \quad \text{or} \quad 16.13 \text{ inches}$$

6. A CYCLOID

The cycloid is an attractive curve and it was given this name in 1661. Many well-known mathematicians have studied the properties of this curve—especially Galileo, Pascal, Bernoulli, and Huygens. If you wish to construct this curve, make a mark on the rim of a tin lid. Roll the lid on a sheet of paper and up against a straight edge. Trace the path of the mark on the paper with a pencil. The resulting curve which repeats itself continuously is a cycloid.

7. FOUR TIMES THE DIAMETER OF THE WHEEL

The mark will trace out a cycloid as the cycle moves forward. At every point where a cycloid touches its base line there is a cusp. The distance from cusp to cusp is πd, but the length of the curve from cusp to cusp is $4d$, where d is the diameter of the generating circle. In this particular case d is the diameter of the bicycle tire.

8. Area of a circle $= \dfrac{\pi}{4}$ (diameter)^2

$\therefore$ area of semicircle $BAC = \dfrac{\pi}{8} BC^2$

But $BC^2 = AB^2 + AC^2$

$\therefore$ area of semicircle $BAC =$ area of semicircle ABM + area of semicircle ACN.

Subtracting the segments common to both sides of this equation: Area of $\triangle$ $ABC =$ area of both lunes.

$\therefore$ area of each lune $= \frac{1}{2}$ area of the triangle ABC. Note that for any right-angled triangle: Area of the triangle $=$ area of both lunes.

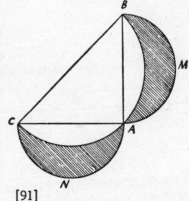

[91]

ANSWERS TO QUIZ NO. 13

1. 16⅝

This problem is to be found in the ancient Rhind papyrus or Ahmes papyrus written more than a thousand years before Christ. The original wording appears strange: "heap, its seventh, its whole, it makes nineteen."

The solution is found thus:

$$x + \frac{x}{7} = 19$$

$$\therefore 8x = 133$$
$$\therefore x = 16\tfrac{5}{8}$$

2. 60

This problem is traditionally attributed to Sun Tsu, who lived in the first century A.D. His method of solving the problem did not involve the usual unknown quantity "x." He did not give us the solution but only the answer—perhaps he guessed. The modern solution is:

Let x = the number of guests.

Equating the number of dishes from the information given,

$$\frac{x}{2} + \frac{x}{3} + \frac{x}{4} = 65$$

$$\therefore 6x + 4x + 3x = 65 \times 12$$
$$\therefore 13x = 65 \times 12$$
$$\therefore x = 60$$

3. 60 YEARS

This is an example of the type of problem which was popular in the fourth century. It is easily solved by letting his age be represented by x years.

$$\text{Then } \frac{x}{4} + \frac{x}{5} + \frac{x}{3} = x - 13$$

$$\therefore 15x + 12x + 20x = 60x - (60 \times 13)$$
$$\therefore x = 60$$

[92]

4. 56 FAT MEN and 40 THIN MEN

To solve this problem, all we need to do is to put the story down in mathematical symbols thus:

$$F - 16 = T \qquad \ldots\ldots 1$$
$$7F - 32 = 9T \qquad \ldots\ldots 2$$

Multiply line 1 by 9:

$$9F - 144 = 9T \qquad \ldots\ldots 3$$

Take line 2 away from line 3:

$$\therefore \ 2F - 112 = 0$$
$$\text{Hence} \quad F = 56$$
$$\text{and} \quad T = 40$$

5. 10 CENTIMETERS—*not too long!*

Translate this problem into mathematical symbols and then there is an easy quadratic equation to be solved.

Let the length of beard be L centimeters.

$$2L + L^2 + 20 = 140$$
$$L^2 + 2L - 120 = 0$$
$$(L + 12)(L - 10) = 0$$
$$\text{Hence} \quad L = 10 \text{ centimeters}$$

6. 90 FEET

Applying the theorem of Pythagoras to both of the right-angled triangles:

$$y^2 = 200^2 + x^2$$
$$y^2 = 150^2 + (250 - x)^2$$
$$\therefore \ 40{,}000 + x^2 =$$
$$22{,}500 + 62{,}500$$
$$- 500x + x^2$$
$$\therefore \ 500x = 45{,}000$$
$$\therefore \ x = 90$$

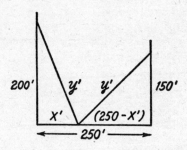

1. *HEXAGON*

The hexagonal shape of the open end of the bee's cell is a familiar sight to some people. In ancient times it was concluded that bees had a certain geometrical sense. The hexagon is one of the few regular shapes that can completely fill the space on a bee frame. Foundation for worker brood is generally made to give comb with about five cells per inch run, thus producing 26 to 29 worker cells per square inch.

2. *TRIANGLE*

The triangle is the only rigid rectilinear figure, and this can readily be proved by joining three, four, or five meccano rods together in a closed figure and applying pressure! This is the reason for the tie bar or tie beam in buildings and gates.

3. *SEMICIRCLE*

The study of old churches is a most fascinating pursuit and frequently one can recognize the period when parts of a church were built. Arches span the openings between chancel and nave, and the heads of windows and doorways. The Norman arches of the twelfth century were semicircular and have a series of characteristic rings.

4. *PARALLELOGRAM*

A parallelogram is a rectangle which has been "pushed over" so that its angles are no longer right angles. Its opposite sides remain parallel and equal. Its diagonals still bisect the parallelogram into two equal triangular areas, but its total area will always be less than the area of the rectangle from which it was formed.

5. *OCTAGON*

The Cross of the Knights of Malta is a white cross of eight points on a black background. If the points are joined, an octagon is formed. A mechanism called a Maltese cross is used in a motion-picture projector to pull the film forward exactly one frame at a time, but this cross has different proportions.

6. SQUARE
It is unusual to find a square in the plant kingdom. It is nevertheless the characteristic feature of the Labiatae family which includes the mints, thyme, sage, ground ivy, the woundworts, the hemp nettles, the dead nettles, and the bugle. They have square stems and the fruit breaks up into four nutlets.

7. PENTAGON
The pentagon is a five-sided plane figure. Five is a common number in flower structure. Many flowers have five sepals and/or five petals (e.g. the buttercup). As soon as one examines the tips of the campanula or Canterbury bell, one immediately recognizes the regular pentagon. Why is this associated with Washington, D.C.?

8. RHOMBUS
This is a square which has been sat on! The rhombus is a parallelogram with all its sides equal, but its angles are not right angles. A square is really a special case of a rhombus.

9. SQUARE
The "great square of Pegasus" is a very obvious shape in the heavens because all four stars at the corners are bright and there are no bright stars within the square. It is clearly seen in the Northern Hemisphere almost due south at midnight during the period of the September equinox.

10. EQUILATERAL TRIANGLE
These three stars are different distances away from the earth. They are well-known stars of the winter's sky. Aratus, a Greek astronomer-poet, writes:

Let Procyon join to Betelgeuse and pass a line afar,
To reach the point where Sirius glows, the most conspicuous
    star,
Then will the eye delighted view a figure fine and vast,
Its span is equilateral, triangular its cast.

[95]

ANSWERS TO QUIZ NO. 15

1. *LIBRA*

The Roman weights and coins were closely linked. About 268 B.C. the new silver denarius was struck, and 72 of these made a libra (or pound). Unfortunately, as the Romans conquered new countries they did not standardize the weights throughout the Empire, so that there were at least eight different Roman pounds.

2. *UNCIA*

This is the Latin for "a twelfth part." It is used with reference to a pound or a foot. The fractional units of these are the ounce and the inch. There are 12 inches in 1 foot but we wonder why there are not 12 but 16 ounces in 1 pound. It is understandable that we all find these units complicated and confusing.

3. *TOWN IN FRANCE*

The troy weight is apparently named after a weight used at the popular fairs held at Troyes in France. Troy weight came into use in England in the thirteenth century, but never ousted the avoirdupois system introduced by the Normans.

4. *CARAT*

The troy ounce was permitted as a legal weight and standard after 1878 only for the sale of precious metals and precious stones. It is obvious that the troy pound and avoirdupois pound both as legal weights would have caused much confusion. The carat weighed originally $3\frac{1}{3}$ grains but is now $3\frac{1}{5}$ grains, or 150 carats make the troy ounce of 480 grains.

5. *HAND*

The "hand" was a linear measure of 3 inches in the sixteenth century but is now standardized as 4 inches. It consists of the width of the palm or hand, and is now used only for the height of horses.

6. ACRE

The Latin word is "ager," meaning a field. Units for land measure varied with time. In medieval England, land was measured in terms of the number of yoke of oxen needed to cultivate the land. Then the acre was the amount that a yoke could plough in a day. This unit is still used.

7. POLE

Years ago, land was ploughed by oxen. The driver's stick or "pole" was used for the oxen and to measure the width of the strip to be ploughed. The pole, rod, or perch is 5½ yards long. A square pole is $\frac{1}{160}$th of an acre, and came into use as a land measurement in England in the fifteenth century.

8. SCRUPLE

The dispenser makes up his prescriptions using the apothecaries' weight, in which a troy ounce is divided into drachms and scruples. The word "scruple" is derived from the Latin word "scrupulus" which is one twenty-fourth of an ounce. The scruple is 20 grains and the drachm is 60 grains. 8 drachms make 1 ounce troy.

9. HENRY I

The name is derived from the Old English word "gierd" or "gyrd," which means a "twig" or "stick." This unit varied in England until 1100, when it was standardized by King Henry I. Throughout the realm it was then taken as the distance from the king's chin or nose to his fingertip.

10. HUNDREDWEIGHT

This is an avoirdupois weight used in measuring heavy things. As its name suggests, it probably originally consisted of 100 pounds, but the custom grew of adding an extra few pounds for perishable goods, so that it varied in various localities from 100 to 120 pounds. It is now standardized as 100 pounds in the United States and 112 pounds in England.

1. TRUE
The pentagram of Pythagoras is the five-pointed star formed by drawing all the diagonals of a regular pentagon and deleting the sides. Pentagrams, heptagrams, and nonograms were considered at a very early date to have magical and mystical properties. The pentagram in particular was used as a symbol of health and happiness. The Pythagoreans (disciples of Pythagoras) used the pentagram as their badge of recognition.

2. FALSE
Zeno was the originator of the paradox of Achilles and the tortoise. Zeno of Elea lived in the fifth century B.C., and he has been associated with many paradoxes of time, space, and motion. Zeno argued that no matter how fast Achilles ran he could never catch the tortoise, for the latter would move a short distance on as Achilles covered the distance between him and where the tortoise was!

3. TRUE
The sensible means of avoiding any confusion between A.M. and P.M. is to use the twenty-four-hour clock. The railway timetables on the continent of Europe leave no possible doubt about the time of the day or night when a train is due to leave a station. Midnight is expressed as 0000 hours and every time during the day is expressed in four digits. For instance, 9:30 A.M. = 0930 hours, and 12 noon = 1200 hours, and 4:15 P.M. = 1615 hours.

4. FALSE
The curve assumed by a chain, rope, or cable when hanging freely between two supports is very much like a parabola, but is really quite different. The curve is called a catenary,

which is derived from the Latin word "catena," a chain. Galileo thought this curve was a parabola, and it was not until late in the seventeenth century that the Bernoullis and Leibnitz discovered the peculiar properties of the catenary.

5. TRUE
The pantograph can be used to copy any figure composed of any combination of straight lines and curved lines. It can be adjusted to cause the copy to be of the same size or to be enlarged or reduced. The instrument consists essentially of a freely-jointed parallelogram of hinged rods. The lengths of the sides of the parallelogram are varied to produce the different sizes of the copy. The original and the copy are in two dimensions and lie in the same plane.

6. FALSE
A histogram is in no way connected with the unit of weight, the gram. It is true that the metric standards are kept at the International Bureau of Weights and Measures at Sèvres near Paris. The histogram is connected with the method of graphical representation of data, and is described elsewhere in this book.

7. TRUE
In architecture the cantilever is a projecting bracket which is used to support a balcony. There are very many examples of its use in buildings of several generations. It is also of great use in the building of bridges. Two cantilevers stretch out from piers on opposite sides and these are joined together by a girder to complete the span. The Quebec Bridge over the St. Lawrence River in Canada has the longest cantilever span in the world—1,800 feet.

[99]

8. TRUE

This is the way in which the term "parameter" is used in mathematics when dealing with the equations of curves and surfaces. It came into use about a hundred years ago. The use of parametric equations frequently simplifies calculations in algebraic geometry and calculus. Parameter means a "side measure." For instance, the co-ordinates of a point on a parabola expressed in terms of one parameter "t" can be written as at^2, $2at$.

ANSWERS TO QUIZ NO. 17

1. *MEDIAN*
This is the definition which is usually given of a median.
The three medians of a triangle are concurrent, and the point
at which they intersect is one third of the way along each
median measured toward the vertex from the mid-point
of the opposite side.

2. *STRAIGHT LINE*
This was the old definition of a straight line. It is doubtful
whether a navigator would agree with this entirely, but as
far as mathematics is concerned this is the meaning usually
attached to it in simple geometry. A better definition would
be "a straight line is one which keeps the same direction
throughout its length."

3. *DIAMETER*
The diameter of a circle is a straight line drawn through the
center and terminated at both ends by the circumference.
Any diameter cuts a circle into two equal parts, each being
called a semicircle. The angle in a semicircle is a right angle.

4. *CHORD*
An important theorem in geometry states that if a tangent be
drawn to a circle and at the point of contact a chord be
drawn, then the angles which the chord makes with this
tangent are equal to the angles in the alternate segments of
the circle.

5. *ASYMPTOTE*
This peculiar word is derived from the Greek language and
first appeared in mathematics in 1656. It was used for the
name of the line to which a curve continually approaches but
does not meet within a finite distance. Frequently an asymp-
tote is called a tangent at infinity.

[101]

6. SECANT

If a straight line cuts any curve at two distinct points, it is called a secant. The beginner in geometry must always differentiate between a secant and a tangent, for the latter, no matter how far it is produced either way, has only one point in common with a curve.

7. PARALLEL OF LATITUDE

This is a small circle drawn through places of the same latitude. It is parallel to the equator and at right angles to the earth's axis or the line joining the North and South Poles. Latitudes are expressed in degrees and minutes on either side north or south of the equator.

8. RADICAL AXIS

This is the locus or path of a point which moves so that the tangents drawn from it to two fixed circles are equal. Actually it is a straight line perpendicular to the line joining the centers of the two circles.

9. DIRECTRIX

The term came into use in 1702. The distance from any point on a conic (ellipse or parabola) to the directrix bears a constant ratio to the distance of the same point from the focus of the conic. For determining the standard equation for a conic the directrix–focus property is generally used.

10. SIMSON or PEDAL LINE

Robert Simson, professor of mathematics at Glasgow University in the eighteenth century, has been honored by having this particular line of a triangle named after him. He made many contributions to mathematics, and most of the English editions of Euclid are based on his work.

[102]

1. PARABOLOID

The surface of a solid obtained by revolving a parabola about its principal
axis is usually called a paraboloid. In
shape it resembles that part of an
acorn which lies outside the cupule.
The volume of a paraboloid is readily found from the equation of the
parabola from which it is generated.

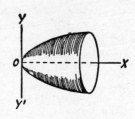

2. ELLIPSOID

This surface is formed by revolving an ellipse about one of its axes
so that all its plane sections are
either circles or ellipses. It is symmetrical about three mutually perpendicular axes. Some birds' eggs
are nearly this shape but most have
a more rounded blunt end. Both
ends of a football are sharper than
an ellipsoid.

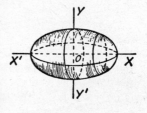

3. CENTROID

The three medians (lines from a vertex to the mid-point of
the opposite side) of a triangle pass through a common point
called the centroid of the
triangle. The term did not
come into general use until
towards the end of the nineteenth century. It is derived
from "center" and "oid." The
center of gravity of a triangular lamina is at this point.

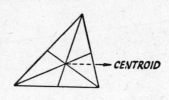

[103]

4. HYPOCYCLOID

The Greek word "hypo" means "under" and the word "epi" means "upon." The hypocycloid is the curve traced out by a mark on the circumference of a circle which rolls on the inside of the circumference of a fixed circle. At every point where the mark touches the fixed circle there is a cusp (a point where two branches of a curve meet). A four-cusped hypocycloid is often called an astroid, and the equation of the curve is less complicated than that of a hypocycloid.

5. EPICYCLOID

The curve was so named by Roemer in the seventeenth century. He showed that cogwheels whose teeth are shaped like an epicycloid curve revolve with minimum friction. If two silver dollars had been used to trace out the curve by rolling one around the other, then the curve formed would be a cardioid. These curves can be seen as caustics by reflection.

6. CISSOID

This curve was discovered by the Greek mathematician Diocles, who flourished as a geometer about 180 B.C. The cissoid was used to accomplish the duplication of a cube, that is, to find the side of a cube whose volume is double that of a given cube. The word is derived from two Greek words meaning "ivy" and "form" which suggests that the curve is "ivy-shaped." Its shape can be seen by plotting the graph of the equation $y^2(2a - x) = x^3$, and Newton devised a method of constructing it mechanically.

1. *THREE*

Early linear measurements were defined in terms of body sizes which, of course, vary from man to man! The earliest English law defining length was made during the reign of Edward II in 1324. It read, "Three barley corns, round and dry, placed end to end, make an inch." Probably this was more of a standard than the width of a man's thumb.

2. *ASTROLABE*

The name is derived from two Greek words meaning "star taking," but the instrument can be used to take the altitude of the sun or moon. By the fifteenth century, quite complicated astrolabes were being used by astronomers, but about 1480 a simpler instrument was made for mariners. The mariner had to know the sun's declination from tables and he could then calculate his latitude from his own observation of the altitude of the midday sun using the astrolabe. This instrument did not give accurate results and in the eighteenth century it was superseded by Hadley's sextant. The maximum angle which could be measured by Hadley's sextant was 90°, hence it was sometimes called a quadrant.

3. *NOCTURNAL*

A nocturnal is an instrument which was used for finding the time by night by observing the relative positions of the North Star and the pointers of the Big Dipper. In the British Museum can be seen a nocturnal made by Humfray Cole in 1560. This instrument was not difficult to use, and when the movable arm was adjusted so that the two pointers of the Big Dipper appeared to lie on it, the time could be read off from a time disk graduated in hours and minutes. The nocturnal was replaced by the chronometer at sea, but was used until quite late in the eighteenth century.

4. *HERO OF ALEXANDRIA*

The exact period of Hero's work we do not know, but he belonged to the First Alexandrian School. We do know that

[105]

he was an able mathematician. About 80 B.C., he put engineering and surveying on a more scientific basis. He is credited with the discovery of this formula for the area of a triangle where s = half the sum of the sides of the triangle.

5. PENTACLE
The pentacle looks like two interlaced triangles. The pentacle and the pentagram are the same figure. It was used as a symbol of mystery by the Greeks, and various societies have used the symbol. In the Middle Ages, many people thought the pentacle had the power to keep away evil spirits.

6. PLANIMETER
This instrument is used for mechanically measuring the area of an irregular plane figure. The hatchet planimeter is probably the simplest type and the wheel and disk, or Amsler type, is the most common.

7. FRUSTUM
Frustum means "a piece broken off." It refers to that portion of a regular solid left after cutting off the upper part by a plane parallel to the base, but it can also be used to describe the portion intercepted between any two planes.

8. APPROXIMATING AN AREA
The area is divided into any even number of parallel strips of equal breadth.

9. CUBE
A hexahedron is a solid figure which has six faces, so that the regular hexahedron is a cube, for it has six equal faces.

10. ELEVEN SECONDS
A "strike" occurs instantaneously and then there is a "rest" before the next strike. Hence, to strike six times requires five rests and this takes five seconds. There are eleven rests when the clock strikes "twelve."

ANSWERS TO QUIZ NO. 20

1. PYTHAGORAS *"to say 'graph'"*

He is one of the best-known mathematicians of the ancient world. He lived in the sixth century B.C. After his studies, he went to Crotona, a Greek city in Italy, and there founded a "school" of wealthy men. Pythagoras included geometry as an essential part of education. Many of the discoveries of his pupils were attributed to Pythagoras. There is no doubt that this Greek philosopher had a great influence. The Pythagoreans were dispersed for political reasons, and Pythagoras died in exile.

2. PLATO *"pal to"*

He was born in 429 B.C., became a pupil of Socrates and, after spending much time in travel, returned to Athens to form a school for students called the "Academy." He was a philosopher and made a study of geometry compulsory before the study of philosophy. It is said that over the entrance to the school was inscribed, "Let no one ignorant of geometry enter." Plato is not associated with any well-known theorems or proofs but he introduced system and method into mathematics.

3. LEONARDO *"rod alone"*

Leonardo of Pisa, or Leonardo Fibonacci, after traveling in Arabia, returned to Italy and in 1202 published a book explaining the Arabic number system. Obviously this was much more practical than the Roman system, with the result that it came into general use among merchants. Leonardo must be given the credit for this achievement. Algebra, geometry, and trigonometry were also dealt with by this mathematician from Pisa.

4. GALILEO *"a gel oil"*

He is sometimes known as the father of dynamics. Galileo was born in Pisa in 1564. The incident referred to in the question took place in the cathedral of Pisa. He timed the

[107]

swings of a large hanging lamp and noticed that no matter whether the oscillations were large or small, the time of swing was the same. He did the timing by using his pulse.

5. PASCAL *"a clasp"*

Blaise Pascal was a contemporary of Descartes. When quite young he displayed a natural aptitude for geometry and very quickly mastered Euclid. He made his reputation both as a philosopher and as a mathematician. He is responsible for much of the original work on the theory of probability as well as the properties of the cycloid. When nineteen years of age he made the first calculating machine, and in 1649 was given the royal right to manufacture the "machines à calculer."

6. BARROW *"rob war"*

Isaac Barrow, who lived from 1630 to 1677, in England, was a much-traveled man. He was educated at Charterhouse, Felstead, and Cambridge. He was elected a Fellow of Trinity College, Cambridge, in 1649. He held many positions of importance in the academic world—professor of Greek at Cambridge, professor of geometry in Gresham College, a Fellow of the Royal Society. He was the first Lucasian Professor of Mathematics at Cambridge but later resigned this post in favor of his brilliant pupil, Isaac Newton, whose superiority he recognized. Barrow became master of Trinity College, and later was elected vice-chancellor of the University of Cambridge. In 1660, he was ordained a clergyman, and continued to urge his conviction that God was ever present and eternal and that there was a divine nature about space and time which accounted for the certainty of mathematics.

7. NEWTON *"not new"*

Isaac Newton, who lived from 1642 to 1727, is the greatest of all English mathematicians. In an attempt to assess his achievements it seems to be impossible to exaggerate! Many mathematicians of other nations maintain that he was the greatest genius of all time.

8. ARGAND *"rag and"*

Jean Robert Argand wrote on the graphic representation of $\sqrt{-1}$, but his work did not at first attract attention. The Argand diagram, which provides a frame of reference for graphing complex numbers, honors this French mathematician.

1. *NOW*

$$\text{Lunn's age} = 2 \times \text{son's age}$$
$$\text{or} \quad L = 2S$$
$$\text{and also} \quad (L - 25) = 3 \times (S - 25)$$
$$\text{or} \quad L = 3S - 75 + 25$$
$$\text{or} \quad 2S = 3S - 50$$
$$\text{or} \quad S = 50$$
$$\text{and} \quad L = 100$$

Thus we drink Grandpa's health just now—he is one hundred years old—a centenarian.

2. 42 *or* 21

This problem admits of two solutions, according to the way you interpret the wording of the limerick.

(a)
$$\frac{B - 1 + 3}{4} = 1 + 10$$
$$\text{or} \quad B + 2 = 44$$
$$\text{or} \quad B = 42$$

(b)
$$\frac{(B - 1) + (B + 3)}{4} = 1 + 10$$
$$\text{or} \quad 2B + 2 = 44$$
$$\text{or} \quad B = 21$$

What a lucky young lady! More than enough boy friends for her to have a different one on each day of the month!

3. 5,000

Clive had in his hive one queen who laid the eggs and headed the colony, as many as 95,000 workers (daughters of the queen—undeveloped females), and only 5,000 drones (sons of the queen—fully developed males).

[110]

Workers + drones = 10^5 = 100,000
But 1 bee in every 20 bees was a drone

Number of drones = $\frac{1}{20} \times 100,000$

$= 5,000$

4. 4
Let H be the number of hits scored by Nero. Then, writing the information given in the form of an equation, we obtain:

$$H^2 - H - 12 = 0$$
$$\therefore \quad (H - 4)(H + 3) = 0$$
$$\therefore \quad H = 4 \quad \text{or} \quad -3$$

Clearly the answer needed here is 4 but what is the significance of the -3?

5. 2 OUNCES
The solution of this problem is similar to that of the last one.
Let the weight of the ball be x ounces.
The equation obtained is another quadratic thus:

$$2x + x^2 - 8 = 0$$
$$\therefore \quad (x - 2)(x + 4) = 0$$
$$\therefore \quad x = 2 \quad \text{or} \quad -4$$

1. Q.E.D.

This is the abbreviation for "Quod erat demonstrandum," a Latin phrase which can be translated as in the question. This dates back to the time when all mathematics books were written in Latin. In the geometry books today these letters appear at the end of each theorem. Placed at the end of each problem are the letters Q.E.F., standing for "Quod erat faciendum," which means "Which was to be done."

2. $\cos \theta$

All trigonometrical ratios of angles are abbreviated. These are written $\sin \theta$, $\cos \theta$, $\tan \theta$, $\operatorname{cosec} \theta$, $\sec \theta$, $\cot \theta$. The cosine of an angle is the ratio of the base to the hypotenuse of the right-angled triangle. Some people remember these different ratios by means of this mnemonic "Some People Have Curly Brown Hair, Till Painted Black."

3. $f(x)$

As an example, $x^2 + 2x - 7$ depends for its value on the value given to x, and it is therefore called a function of x and is written $f(x)$. If $f(x) = 3x - 5$, and if x is given the value 3, then $f(3) = 3 \cdot 3 - 5 = 9 - 5 = 4$.

4. $\int 16x^3 dx$

$\int$ is called the operator and shows that the operation of integration is to take place on $16x^3$, and dx makes it clear that the integration is to be with respect to x. $\int 16x^3 dx = 4x^4$. If $4x^4$ were differentiated we should obtain $16x^3$. Integration and differentiation are two processes closely related to each other.

5. L.C.M.

This is the mathematical shorthand for "lowest common multiple." Thus the L.C.M. of 4, 8, and 12 is 24, because 24 is the smallest whole number into which 4, 8, and 12 will divide exactly.

6. *sinh x*

The functions $\frac{1}{2}\,(e^x - e^{-x})$ and $\frac{1}{2}\,(e^x + e^{-x})$ possess properties analogous to sin x and cos x. These functions are therefore defined as "hyperbolic sine" and "hyperbolic cosine" of x. Sinh $x = \frac{1}{2}\,(e^x - e^{-x})$, and cosh $x = \frac{1}{2}\,(e^x + e^{-x})$. Just as $\sin^2 x + \cos^2 x = 1$, so $\cosh^2 x - \sinh^2 x = 1$.

7. *i*

If $x^2 = -1$, then we can find no real number to satisfy the equation. The Swiss mathematician, Euler, introduced the symbol i for $\sqrt{-1}$. The symbol is used when dealing with "complex numbers." It is also essential when studying both the theory of air-flow patterns and alternating currents.

8. *G.C.D.*

This is the abbreviation for "greatest common divisor." Thus 3 is the G.C.D. of 6, 9, and 12. It is usual to find the G.C.D. by writing down the prime factors of each number and noting those that are common to all.

9. $\frac{dy}{dx}$

This is the differential coefficient of y with respect to x, or the first derivative of y with respect to x. It can also be considered to be the gradient of the tangent to the graph of y plotted against x. If $\frac{dy}{dx}$ is constant at different points along the graph, then the graph is a straight line; if $\frac{dy}{dx}$ varies, then the graph is a curve.

10. *e*

The eccentricity of a conic is the ratio between the distance of a point on the curve from the focus and the distance of the same point from the directrix. The following values for e are always true: $e < 1$ for the elapse, $e = 1$ for the parabola, and $e > 1$ for the hyperbola.

[113]

ANSWERS TO QUIZ NO. 23

1. ASTRONOMY

The science of star-arranging or classifying seems to narrow down astronomy, but the subject matter of astronomical science includes all the matter of the universe that lies outside the limit of the earth's atmosphere. Such a vast subject is subdivided, and the mathematician can come into his own with celestial mechanics.

2. ARITHMETIC

The word "arithmetic" is derived from a Greek word meaning "the art of counting," which in turn comes from another Greek word meaning "number." The scientific treatment of numbers with the Greeks had special reference to ratio, proportion, and the theory of numbers. Before algebra was considered to be the subject of a separate study, arithmetic embraced all number knowledge, including all that we should now term algebra. Perhaps a more modern meaning of the word "arithmetic" would be "the art of numerical calculation and its immediate applications."

3. TRIGONOMETRY

The subject of trigonometry deals with the measurement of the sides and angles of a triangle and with functions of their angles. Early theories in astronomy were halted until the invention of trigonometry in the second century B.C. Hipparchus, an eminent Greek astronomer, can be called the father of trigonometry.

4. GEOMETRY

The subject first developed was land surveying, and this took place particularly in Egypt. When the Nile overflowed its banks and flooded the fields, the landmarks were removed. It became essential then to develop a system of surveying. Hence the derivation of the word "I measure the earth." In this way we have the origins of the subject. The Greeks were quick to develop the subject in a more abstract way.

It is true to say that all the leading mathematicians have been interested in geometry.

5. ALGEBRA
The first Arabian mathematician of note is Alkarismi (ninth century) and he wrote an algebra book entitled *Al-jebr we' l mukabala*. In this book he shows how to treat equations by (a) taking quantities from one side to the other, (b) uniting similar terms into one term. From the word "Al-jebr" our word "algebra" is derived. In the twelfth century mathematical works in Arabic were translated into Latin, and then Europe possessed the tools for further progress.

6. MATHEMATICS
The word "mathematics" came into use in the late sixteenth century and it is derived from a Greek word meaning "something learned" or "science." Broadly speaking, it is divided into two sections: (1) pure mathematics, which is the abstract science of space and number, (2) applied mathematics, which, as its name implies, deals with the application of mathematics to all branches of science and engineering.

7. STATICS
In modern use this branch of applied mathematics is concerned with the action of forces in producing equilibrium or rest, in contrast to dynamics, which deals with the action of forces in producing motion.

8. CALCULUS
This word came into use in 1672, and is derived from the fact that calculations were made on an abacus with the aid of small stones or pebbles. There will always be arguments as to whether Newton or Leibnitz invented calculus. It is now accepted that Newton was the first inventor and Leibnitz did not "borrow" his ideas! We actually use the notation originated by Leibnitz.

ANSWERS TO QUIZ NO. 24

1. ABACUS
About two thousand years ago merchants would count by setting out pebbles in grooves of sand. The Romans set out pebbles in grooves in a metal plate. Such were the early models of the calculating frame. The type of abacus with sliding balls on wires came into use in England in the late seventeenth century.

2. CLEPSYDRA
This name is derived from two Greek words, and the device was designed by the ancients to measure time by the discharge of water. The water clock was used by Egyptians more than three thousand years ago, but better models were made by the Greeks about 400 B.C. Such a device enabled people to tell the time at night.

3. TALLY
The tally or the tally stick was a piece of wood scored across with notches for the items of an account. The size of the notch varied from the small pence notch to larger ones for a shilling and a pound and so on. The stick was then split into halves of which each party kept one. Hence we see the derivation of the term "our accounts tally."

4. FRENCH CURVE
Accurate curves are essential in engineering drawing. Circles are drawn with special compasses. For other curves a French curve is used, a portion of this being selected to fit the curve to be drawn.

5. SEXTANT
This astronomical instrument includes a graduated arc equal to a sixth part of a circle, and it is used in particular for observing altitudes of celestial bodies in order to fix a latitude at sea. The instrument was first described to the Royal Society in England by Robert Hooke in 1667.

6. THEODOLITE

The origin of this useful instrument is not known. It was originally used for measuring angles in a horizontal plane. The essential parts are a horizontal plane and two "sights" such as are found in an alidade. With the telescope came better models, and further improvements came with levels, verniers, micrometers, and a vertical scale.

7. SLIDE RULE

John Gunter invented a slide rule in 1620. A logarithmic scale was marked on a single line and by means of dividers, multiplications and divisions could be worked out by adding and subtracting lengths on the scale. Two Gunter's scales arranged to move side by side were more practical and were soon incorporated in the design.

8. ADDING MACHINE

The adding machine was invented by Pascal in 1642 when he was nineteen years of age and he received royal permission to be the only maker in France. The adding machine was modified by Leibnitz to include multiplication. Improvements were made in the nineteenth century and difference machines soon followed.

9. CROSS HEAD

This consists of two pieces of wood at right angles to each other. It is used to ensure that accurate offsets are made from a survey line when making a field survey.

10. SCREW GAUGE

This instrument enables one to measure with great accuracy very small lengths placed between metal jaws. One of the jaws is the end of a fine screw which has a pitch of 1 mm. A scale drawn around the head of the screw is divided into 100 parts so that a difference of 0.001 cm. can easily be read.

ANSWERS TO QUIZ NO. 25

1. CURSOR

The cursor is defined as a part of a mathematical instrument, which slides backwards and forwards. Newton suggested a cursor or runner should be used on a slide rule but his idea was not taken up for a hundred years. It is obvious when one is trying to read the graduations on two different scales that some convenient straight line is necessary, and the cursor is now an essential part of any slide rule.

2. MUSICAL SCALE

From our childhood we are accustomed to "do, re, mi, fa, sol, la, ti, do," which is known as the diatonic scale. Other scales are to be found in Turkish and Persian music! The frequencies of the eight notes of a diatonic scale are in the following ratio:

$$\frac{1}{1} : \frac{9}{8} : \frac{5}{4} : \frac{4}{3} : \frac{3}{2} : \frac{5}{3} : \frac{15}{8} : \frac{2}{1},$$ which is the same as

24 : 27 : 30 : 32 : 36 : 40 : 45 : 48, giving the notes
C   D   E   F   G   A   B   C'

3. TWICE

If you doubt this try it by marking the rolling half-dollar with a small spot on its circumference. The answer usually given is "one," and the false reason supporting the answer is that the circumferences of H and D are the same. However, on doing the experiment you will notice how the spot traces out a special curve.

4. CYCLOID

This is the simplest member of the class of curves known as roulettes. It was not known before the fifteenth century and not seriously studied until the seventeenth century. So many brilliant mathematicians like Descartes, Pascal, Leibnitz, the Bernoullis, and others have investigated the properties of the cycloid that it was sometimes named the "Helen of Geometers." It is a matter of opinion whether the curve, like Helen of Troy, is of surpassing beauty.

[118]

5. DESARGUES' ANSWER

OA, OB, and OC are three concurrent straight lines with X, Y, and Z any points on each respectively. Join CA and ZX and produce them to meet in R. Similarly, join AB and XY to meet in S, and CB and ZY to meet in T. Gérard Desargues, a seventeenth-century French engineer, proved that R, T, and S always lie on a straight line. Hence, if you plant your ten tulips in the positions of the ten points A, B, C, X, Y, Z, R, T, S, and O, they will be in ten rows of three.

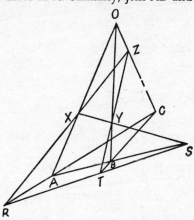

6. ABOUT THREE INCHES

The needle moves from the outermost groove to the innermost groove in an arc whose radius is the length of the pickup arm.

7. MR. PHILLIPS

Mr. Henry will earn during the first year $3,000.

Mr. Phillips will earn during the first year $1,500 + $1,500 + $300 = $3,300.

Mr. Henry will earn during the second year $3,000 + $600 = $3,600.

Mr. Phillips will earn during the second year $1,800 + $300 + $2,100 + $300 = $4,500.

Hence it is clear that Mr. Phillips will earn a sum of $1,200 more than Mr. Henry by the time they have both worked two years. As the years go by Mr. Phillips will earn increasingly more than Mr. Henry.

[119]

1. $\frac{2}{15}$

The two relevant equations are: $B + C = \dfrac{2}{5}$

and $\quad B = 2C'$

From these we derive that $3C = \dfrac{2}{5}$ or $C = \dfrac{2}{15}$

2. $20\frac{20}{87}$ yards.

A runs 1,760 yards while B runs 1,740 yards and C runs 1,720 yards.

∴ B runs 1,740 yards while C runs 1,720 yards

∴ B runs 1,760 yards while C runs $\dfrac{1,720 \times 1,760}{1,740}$ yards

or $1,739\frac{67}{87}$ yards

∴ B beats C by $20\frac{20}{87}$ yards in a mile

3. 60 *DAYS*

$(A + B)$ do $\frac{1}{10}$ of the work in one day

$(A + C)$ do $\frac{1}{12}$ of the work in one day

$(B + C)$ do $\frac{1}{20}$ of the work in one day

Add these all together and we have:

∴ $2(A + B + C)$ do $\left(\frac{1}{10} + \frac{1}{12} + \frac{1}{20}\right)$ of the work in one day

or $(A + B + C)$ do $\frac{7}{60}$ of the work in one day

but $(A + B)$ do $\frac{1}{10}$ of the work in one day

∴ C alone does $\left(\frac{7}{60} - \frac{1}{10}\right)$ of the work in one day

or C alone does $\frac{1}{60}$ of the work in one day

∴ C can do all the work in sixty days

4. 18

A scores 50 while B scores 40

B scores 50 while C scores 40

$\therefore$ B scores 40 while C scores $\dfrac{40 \times 40}{50}$

or 32

$\therefore$ A scores 50 while B scores 40 and while C scores 32

$\therefore$ A can give 18 points to C

5. $500

The profit should be shared in the following proportions between A, B, and C: 1,875 : 1,500 : 1,250

or   15 :   12 :   10

or $\dfrac{15}{37}$: $\dfrac{12}{37}$: $\dfrac{10}{37}$

$\therefore$ C's share of the profit should be $\dfrac{\$10 \times 1,850}{37}$

or $500

6. $1\frac{11}{19}$ HOURS

In one hour A fills $\frac{1}{2}$, B fills $\frac{1}{3}$, and C empties $\frac{1}{5}$ of the tank.

$\therefore$ with all the pipes working $\left(\frac{1}{2} + \frac{1}{3} - \frac{1}{5}\right)$ or $\left(\frac{19}{30}\right)$ of the tank is filled in one hour

$\therefore$ with all the pipes working $\left(\dfrac{30}{30}\right)$ of the tank is filled

in $\left(\dfrac{30}{19}\right)$ hours.

[121]

1. $X = 9$, $Y = 1$, $Z = 8$

Examine the units column. $X + Y + Z = 10 + Z$, which means that $X + Y = 10$. Examine the tens column. $X + Y + Z + 1$ (from the units column) $= 10 + X$, which means that $Y + Z = 9$. Examine the ten-thousands column and the equation obtained is: $X + Y + Z + 1$ (from the thousands column) $= 10Y + X$. Substitute $Y + Z = 9$, and the equation then becomes: $10 = 10Y$, or $Y = 1$. From this it follows that $Z = 8$, and $X = 9$.

$$\begin{array}{r} 9999 \\ 1111 \\ 8888 \\ \hline 19998 \\ \hline \end{array}$$

2. $X = 5$, $N = 2$, $P = 1$, $S = 0$, $R = 6$, and $Z = 3$

Examine the second row of multiplication by X. X times X gives another X. $\therefore$ X must be 0, or 1, or 5, or 6. $X \neq 0$ because the product is not XXX. $X \neq 1$ because the product is not PNX. $X \neq 6$ because there is no product of NX by X which will give ? NX. $\therefore$ X = 5, because $25 \times 5 = 125$ or $75 \times 5 = 375$. $\therefore$ N = 2 or 7. Examine the first row of multiplication by N. $N \neq 7$ because the product is not one of four figures. $\therefore$ N = 2. The remainder of the unknown letters follow fairly easily from this stage.

$$\begin{array}{r} 125 \\ 25 \\ \hline 625 \\ 250 \\ \hline 3125 \\ \hline \end{array}$$

3. $P = 3, H = 1, I = 2, L = 5, T = 6$, and $S = 0$.

From the first division it is clear that $H = 1$. From the third division L times L gives another L in the units place. $\therefore$ L must be 0, or 1, or 5, or 6. $L \neq 0$ because the product is not LL. $L \neq 1$ because the product is not IL, and in any case $H = 1$. $L \neq 6$ because whatever value is given to I, between 2 and 9, the product of 6 times I6 will not be 1I6. $\therefore L = 5$. $\therefore$ HIL now reads 1I5. Examine the third division again. 5 times I5 is <200. $\therefore$ I must be 2 or 3. Examine the second division. In order to produce the product 5S, 2 times 25 would be possible and 3 times 35 impossible. $\therefore I = 2$ and $S = 0$. The other letters are easily decoded from this stage.

$$
\begin{array}{r}
125 \\
25\overline{)3125} \\
25 \\
\hline
62 \\
50 \\
\hline
125 \\
125 \\
\hline
\cdots
\end{array}
$$

4. $A = 3, E = 9, F = 6, H = 0, L = 7, N = 1, P = 2$, and $Y = 5$

By a similar argument as in answer 3, Y must be 5 or 6. From the second subtraction in the units column $L - H = L$, $\therefore H = 0$. From the second division, in order to produce 0 in the units column, F must be even if $Y = 5$ and F must be 5 if $Y = 6$. From the second subtraction in the tens column $P - N = N$. $\therefore P = 2N$. $\therefore N \not> 4$, and P is even. From the first subtraction in the units column $L - Y = P$. As Y is 5 or 6 the only values P can have must be 2 or 4. $\therefore$ N must be 1 or 2. Now consider F and Y again, trying

[123]

$F = 5$ and $Y = 6$ in the second division. An impossible result is obtained so that F must be even and $Y = 5$. Examine the first subtraction. $E - L = P$ and $L - Y = P$. By adding these $E - Y = 2P$. $\therefore E = 2P + 5$. $\therefore P = 2$. $\therefore L = 7$. This is the key and the remaining letters follow quickly.

$$
\begin{array}{r}
565 \\
\overline{35)19775} \\
175 \\
\overline{} \\
227 \\
210 \\
\overline{} \\
175 \\
175 \\
\overline{} \\
\cdots
\end{array}
$$

1. *TYCHO BRAHE*

He was the famous sixteenth-century Danish astronomer. It was as a result of his systematic observations of the sun and the planets that Kepler discovered his three laws. Mathematically, Tycho Brahe's conception of the solar system was the same as that of Copernicus. His approximation of π was 3.1409, but it is not known how Tycho Brahe arrived at this curious value. He perfected the art of astronomical observation before the advent of the telescope, and included in his instruments was a large mural quadrant. The size of the quadrant was so great that it took twenty men to transport it to its operating site!

2. *THALES*

He was the founder of the earliest Greek school of mathematics and philosophy. He was born at Miletus in Asia Minor about 640 B.C., went as a merchant to Egypt, and there he studied astronomy and geometry. Tradition says that he was the first to find the height of a pyramid by means of the shadow-stick method.

3. *ARCHIMEDES*

This famous Greek mathematician and inventor was born at Syracuse in Sicily. His greatest work was done in geometry. He was killed by a Roman soldier when Syracuse was captured in 212 B.C., but he was given an honorable burial. His tomb was marked by a sphere inscribed in a cylinder, for Archimedes considered that his greatest discovery was that both the volume and the surface area of the sphere are two-thirds of the corresponding measurements of the circumscribing cylinder.

4. *DESCARTES*

René Descartes is regarded by some as the father of modern philosophy. In mathematics also this brilliant seventeenth-century scholar achieved lasting fame. He published in 1637 the first book of co-ordinate geometry. The great advance made by Descartes was the plotting of points to form a curve by using two axes at right angles.

5. PTOLEMY

We know little about Claudius Ptolemaeus except that he lived in the second century A.D., in Alexandria, and that he was an astronomer and a mathematician. His great book was translated into Arabic and it is from the Arabic that we get its title *Almagest*. The work is divided into thirteen books, and it is obvious to anyone studying these books that Ptolemy was a great geometrician. The suggested crest is the figure for Ptolemy's theorem found in geometry books.

6. GUNTER

Edmund Gunter was born in England in 1581. He was appointed professor of astronomy at Gresham College and became a noted inventor and able mathematician. Gunter's chain, devised in 1610, is used in land surveying. It is 66 feet or 22 yards in length and is divided into 100 links. It has an obvious advantage as far as English square measure is concerned for 10 square chains equal 1 acre.

7. KEPLER

His three laws have given this German astronomer great renown. Johann Kepler became associated with Tycho Brahe and when the latter died Kepler had access to his many astronomical observations. From these he discovered the laws which govern the motions of the planets. He found out that the movements of Mars could only be accounted for if the sun was at the focus of an ellipse and the path taken by the planet was the ellipse itself.

[126]

8. GALILEO

Galileo Galilei was born in the Italian town of Pisa, which is noted for its leaning tower. This well-known scientist and astronomer was appointed professor of mathematics at Pisa in 1589. Aristotle had taught that heavy bodies fall faster than lighter ones, but Galileo proved by experiment that this is not true. The public exhibition from the leaning tower of Pisa is now regarded as legendary, but it makes a good story!

1. *OGEE*

The ogee arch is constructed as in the diagram. This type of arch became especially popular in the fourteenth century in Italy, but owing to its weakness it could only be used in the windows of a church. The points C in this and the other diagrams represent the centers of curvature of sections of the arch.

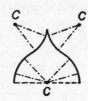

2. *LANCET*

The lancet arch is composed of two arcs. The centers of these arcs are situated on the springing line produced and outside the arch itself. Lancet windows were popular in the early thirteenth century when two or more of them were placed close together to secure as much light as possible.

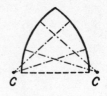

3. *MOORISH*

This is a pointed arch consisting of two arcs but it differs from the lancet type in that the centers for describing the arcs lie within the opening of the arch itself.

This architectural feature appeared for the first time in the ninth century in a mosque in Cairo, and soon became an emblem of the Mohammedan faith. The Moors in Northwest Africa did not use it.

4. *RAMPANT*

The rampant arch has, as a special feature, springing points at different levels, but this is essential because the arch is used to support a flight of steps that may be solid.

5. EQUILATERAL

This type is constructed on an equilateral triangle, as in the diagram. This arch was popular in the late thirteenth century, and you will notice that it is much wider than the lancet type. This arch is sometimes called the Early English or pointed arch.

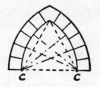

6. SEGMENTAL

As its name implies, the segmental arch is formed from the segment of a circle. The Romans are really responsible for the introduction of arched construction work. In the segmental arch we have a wonderful example of how bricks can be arranged in a curved structure so that they give mutual support to one another.

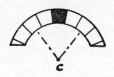

7. HORSESHOE

The arc of the horseshoe arch is greater than a semicircle, with its center above the springing line of the arch. This arch was widely employed in Moorish architecture. It was the horseshoe arch that the Moors brought into Spain, where they built, among other things, the Great Mosque at Cordova. The arch was copied by the French in the south, and ultimately a few doorways and windows were decorated by it in England toward the end of the twelfth century.

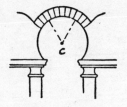

[129]

8. FOUR-CENTERED (TUDOR)

This arch was first made in the building revival that came
with the Tudors and so it is sometimes called the Tudor
arch. It is formed by using the arcs of
four circles, and is a much stronger and
wider arch than the ogee arch. The
builders of churches of this period had
to provide large east windows so that
the four-centered arch was the perfect
answer.

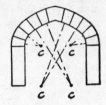

9. NORMAN

The so-called Norman period dates from 1066 to 1189. The
Normans were great builders, and after the conquest of
Britain many church buildings were erected there. It is a
common mistake to think that a
semicircular arch in a church is
necessarily of Norman construction. The pillars, moldings, or
ornamentations must be exam-
ined. The plain zigzag ornamentation called the chevron is frequently found on the Norman
arch. Saxon arches were semicircular but their moldings, if present, were very plain.

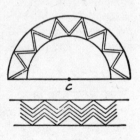

1. *T*
The line which touches a circle is called a tangent.

2. *T*
A score is 20. The twentieth letter of the alphabet is T.

3. *O*
0 times 123 is 0, which looks like O.

4. *A*
The abbreviation for acceleration is the small letter *a*. Acceleration, like velocity, is a vector quantity because it must have direction as well as magnitude.

5. *H*
Horsepower is a measure of how fast work is being done. The unit in the United States and England is known as the Watts horsepower and its value is 550 foot-pounds per second.

6. *M*
The standard unit of length in the metric system is the meter.

7. *D*
A decagon is a plane figure or surface having ten sides and ten angles.

8. *U*
This is a quadratic equation.

9. *R*
R represents the rate of interest per cent per annum.

Hence the letters are in the order TTOAHMDUR and these when rearranged form the name DARTMOUTH.

ANSWERS TO QUIZ NO. 31

ACROSS

1. *FIBONACCI.* Leonardo Fibonacci (or the son of Bonacci or filius Bonacci) was the first man to bring and explain the Arabic system of numbers to Europe. He was a great champion in the popular mathematical tournaments of the thirteenth century. He wrote both an arithmetic and a geometry book.

2. *CURE.* During the year 1658 it is said that Pascal was suffering from toothache when the idea occurred to him of writing the geometry of the cycloid. He began the work and the pain disappeared, so he continued and completed it in eight days.

3. *BRAC.* Brackets has eight letters—omit the last four.

4. *YU.* YOU minus O equals YU!

5. *OD.* These are ODD numbers!

6. *AS.* The recognized abbreviation for anna.

7. *LILEO.* Possibly a twice-beheaded Galileo would have been a simpler clue.

8. *BA.* Bachelor of Arts, or B.A., which is the degree given by many colleges, although one might expect a B.Sc. in mathematics.

9. *IE.* This is the abbreviation for "id est," the Latin for "that is."

10. *LB.* The abbreviation for the pound derived from the Latin "libra," meaning "pound."

11. *RR*. Robert Recorde, who wrote *The Whetstone of Witte*, an algebra book, in 1557.

12. *KG*. 1,000 grams is a kilogram, and is the commonest unit used in France. It is equivalent to 2.2 pounds weight.

13. *ARCHIMÈDE*. This is the way in which "Archimedes" is spelled in French.

DOWN

3. *BOOLE*. A nineteenth-century English mathematician who for many years was professor of mathematics at Cork. He wrote many books. His treatise on the calculus of *Finite Differences* is a classic.

14. *IVORY*. Sir James Ivory was really a self-taught mathematician who lectured at the Royal Military College with great success. Much of his work was published in various journals.

15. *OHM*

16. *ARC*

17. *CHORD*

18. *MACLAURIN*. As a boy he was a good geometer and as a young man he occupied a chair of mathematics at Aberdeen. Later he was recommended by Newton to Edinburgh. He wrote a treatise on *Fluxions* in which the method of distinguishing between "max and min" points on a curve was first given. Maclaurin's theorem is of the greatest use in expanding certain functions.

[133]

19. *RECTANGLE*

20. *EULER.* He was considered a genius even by the Bernoullis. Curiously, he was appointed to a chair of mathematics in Russia. He was a prolific writer, not only on mathematical subjects but also on scientific and medical subjects.

21. *TALLY.* Counting was often done by cutting notches on a piece of wood. The word is derived from the French word "tailler," meaning "to cut."

22. *SIR.* Isaac Newton was knighted on April 16th, 1705.

23. *MPH.* Miles per hour.

24. *CAM.* The cam shaft of a four-stroke internal-combustion engine moves the intake and exhaust valves.

25. *LCD.*

1. *AN AID FOR READING A NUMBER*

Tonstall, who became Bishop of London during the early part of the sixteenth century, gave some interesting information about the "new" Hindu-Arabic number system and arithmetic in England. In his book *De Arte Supputandi* he shows place value by means of dots or points above the figures. We have become used to commas being placed in such a way as to group the figures in threes. Hence 567342452 is now written as 567,342,452.

2. *PREPARATION TO FIND THE SQUARE ROOT*

Tonstall in his arithmetic book includes a method of finding the square root of a number. Just as we mark off in pairs from the decimal point, he does the same thing but shows it by means of the superior point placed over the first figure in each group. Thus 98525476 is the same as $\overline{98}\,\overline{52}\,\overline{54}\,\overline{76}$ or as 98'52'54'76.

3. *MULTIPLICATION OF 9 BY 6*

Robert Recorde's arithmetic book *The Grounde of Artes* was the arithmetic book of the century 1543-1643. In this book he gives the rule for multiplying 6 by 6 up to 9 by 9. In the question we have to multiply 9 by 6. A large cross is drawn and the 9 and the 6 are written on the left-hand side. Subtract each from 10 to give 1 and 4 respectively. Now multiply the two differences together, 1 by 4 = 4 and this is the units figure in the answer. Subtract either the 4 from the 9 or 1 from the 6 (that is why the cross is drawn) and then we have the tens figure which is 5. So the answer is 54. We wonder if it isn't easier to learn your tables.

4. *DIVISION BY 10*

Recorde suggested a vertical line to show the division when a number was divided by ten.

5. DIVISION OF VULGAR FRACTIONS

Another process shown by Recorde is the multiplication and division of fractions. In division, which we choose in this question, there is no rule about turning the divisor upside down and multiplying. The division is carried out by cross-multiplying. Cross-multiply the 1 and 4 for the numerator and the 3 and 3 for the denominator.

6. WRITING A DECIMAL

83 4' 2" 5''' is the same as 83.425. A commercial arithmetic was published by Francesco Pellos in Turin in 1492, and in it he unwittingly used the familiar decimal point. Later writers, however, used a bar to represent the decimal point. John Napier, the inventor of logarithms, used the style as in the question but later changed this. There must have been at least a dozen methods of writing decimals and today $23 \cdot 45$; 23.45; and 23,45 are used in different countries.

7. MULTIPLICATION TO FIND AN AREA

This example is taken from John Bonnycastle's book *The Scholar's Guide to Arithmetic,* which was published in 1780. The first line of multiplication is done by 5 feet and the second line by 4 inches. In 1780 the answer would have been given as 19 square feet, $6\frac{2}{3}$ square inches. This is wrong. It should read "19 square feet, 6 'feet-inches,' and 8 square inches," and this makes 19 square feet and 80 square inches—the correct answer. It is the same as the area of a strip 1 foot wide and 19 feet, $6\frac{2}{3}$ inches long.

8. RULE OF THREE *or* PROPORTION

If five books cost 25 dollars, what will eight books cost? The 5 and the 8 being of the same kind (books) are placed on the same side, and the 25 dollars is placed opposite the 5 because they are linked together. The rule as stated by Recorde is to multiply 8 by 25 and divide by 5 and hence the cross line!

1. 1564. *WILLIAM SHAKESPEARE*
There are four digits in the year we are searching for in this story. Suppose we write the year as $ABCD$.
$A = 1$ because we have not yet reached the year 2000. Let us form equations from the facts given in the story:

$$A + D = B \quad \ldots\ldots\ldots 1$$
$$C = B + 1 \quad \ldots\ldots 2$$
$$3D = 2C \quad \ldots\ldots\ldots 3$$

Put in equation 3 a value for D in terms of B from equation 1 and at the same time put a value for C in terms of B from equation 2 and we have:

$$3(B - 1) = 2(B + 1)$$
$$\text{or} \quad B = 5$$
$$\text{Hence} \quad C = 6 \quad \text{and} \quad D = 4$$

Thus our gentleman was born in 1564. Who else could this be but William Shakespeare?

2. *JULIETTE* 16⅔, *or* 30 *METERS, NEARER MON-TREUX. LUCILE MISSES THE BUS EVERY-WHERE*
Draw your own diagram of a straight road and letter the position of the bus B, the point where the sisters left it P, the patch of narcissi where Juliette is J, and the correct meeting point with the bus for Juliette M.
Let us call the distance PM x meters.
Juliette runs MJ meters during the time the bus travels BM meters.

$$\therefore 2 \times MJ = 1 \times BM$$
$$\text{But} \quad MJ = \sqrt{40^2 + x^2} \quad \text{(Pythagoras)}$$
$$\text{and} \quad BM = 70 + x$$
$$\therefore 2\sqrt{40^2 + x^2} = 70 + x$$

[137]

$$\therefore \quad 6{,}400 + 4x^2 = 4{,}900 + 140x + x^2$$
$$\therefore \quad 3x^2 - 140x + 1{,}500 = 0$$
$$(3x - 50)(x - 30) = 0$$
$$x = 16\tfrac{2}{3}, \quad \text{or} \quad x = 30 \text{ meters}$$

Thus Juliette could have run to a point nearer Montreux by $16\frac{2}{3}$, or 30 meters from the point where they left the road for the field, and she would have caught the bus.

Lucile was 41 meters from the road so her equation in the same way is:

$$2\sqrt{41^2 + x^2} = 70 + x$$

which becomes:

$$3x^2 - 140x + 1{,}824 = 0$$

and this equation will not give real roots like Juliette's equation. Thus poor sister Lucile missed the bus and wherever she had run she could never have caught it!

3. TAKE MY CAMEL ALSO AND THEN DIVIDE
When they included the camel belonging to the dervish there were eighteen camels in all. At first the eldest son took his share ($\frac{4}{9}$ of 18), that accounted for eight camels. The second son next took ($\frac{1}{3}$ of 18), or six camels, and finally the youngest one took his share, ($\frac{1}{6}$ of 18), that is three camels. To their surprise one camel was left over.

The dervish mounted the remaining camel, which happened to be his own, smiled and rode away waving friendly greetings.

4. YES
This is a Möbius band called after the German mathematician Möbius, who lived from 1790 to 1868. It is used in the modern branch of geometry called topology.

You can prepare such a surface in this way: Take a long strip of paper, hold one end firmly, twist the other end through 180°, and then seal the two ends together in that position. You have formed a continuous band of paper with a single twist in it. This is the Möbius band. You can convince yourself of its peculiar property by drawing a pencil line along the middle of the band until you return eventually to the starting point. In this way you will see that the band has one side only. You can follow the edge all the way around in the same manner. Just for fun, guess what would happen if you cut along the center pencil line. Try it!

Now guess what would happen if you cut along the center line of the band or bands produced by the first cut. Try this. Is this a surprising result? Examine the twists. How many are there altogether?

You have learned by this time a great deal about surfaces and edges. Take another Möbius band and this time cut lengthwise a third of the way from one edge until you arrive at the starting point. It is even more difficult to think what the result will be! It is probably very different from what you thought.

ANSWERS TO QUIZ NO. 34

1. SEVEN
They are the pyramids, the mausoleum at Halicarnassus, the hanging gardens of Babylon, the temple of Diana at Ephesus, the Colossus of Rhodes, Jupiter's statue by Phidias, and the Pharos at Alexandria.

2. THREE
Three sisters, daughters of Zeus. Their names are Euphrosyne, Aglaia, and Thalia. They were bosom friends of the Muses and were goddesses of beauty and grace, who distributed joy and gentleness. Usually they were pictured as embracing each other, showing that where one is, there also are the others.

3. FOUR
The same as in an American deck of cards. There are le roi de cœur, le roi de pique, le roi de trèfle, and le roi de carreau.

4. SIX HUNDRED AND SIXTY-SIX
The reference is to chapter 13 verse 18 of the Revelation of St. John the Divine. "Let him that hath understanding count the number of the beast . . . and his number is six hundred three score and six." In spite of efforts down the ages to find the beast, it is now accepted that the beast lived in the age when the book was written, and that he was Nero.

5. NINE
There are nine Muses, the nine daughters of Zeus and Mnemosyne. They were supposed to be the goddesses of memory, but later they were identified with the arts and sciences. Their names are Calliope (the chief), Clio, Euterpe, Thalia, Melpomene, Terpsichore, Erato, Polyhymnia, and Urania. Clio is associated with history, Terpsichore with song and dance, and Urania with astronomy. Originally there were only three and they were worshipped on Mount Helicon. Then the number grew to seven, and later to eight. Finally nine became established throughout Greece with the names given here.

[140]

6. TWO

The Two Gentlemen of Verona was first printed in the folio of 1623 as the second of the comedies of William Shakespeare. The names of the two characters in the play are Valentine and Proteus.

7. FOUR

Aristotle taught that there were four elements, namely fire, air, water, and earth. Later a fifth was added called "quintessence," and this was supposed to be in the four other elements and its purpose was to unify them. Shakespeare writes in *Twelfth Night* this sentence, "Does not our life consist of the four elements?" The modern definition of an element is totally different. Today there are one hundred known elements and there is a possibility that more will be discovered.

8. FIVE

This is a quotation from *The Owl* by Tennyson. There are five wits—common sense, imagination, fantasy, estimation, and memory. Common sense is the outcome of the five senses, and it appears that Tennyson had in mind the five senses when he wrote these lines.

9. ONE HUNDRED

When Napoleon escaped from Elba in 1815 he came to the Tuileries. This was on the 20th of March. It was not until one hundred days had elapsed, on the 28th of June, that Louis XVIII was restored as monarch. The quotation in the question is taken from the address of the Prefect of Paris which he made to the returning king. One very important event which took place during the hundred days was the Battle of Waterloo.

10. SEPT

This proverb means "You must think before you speak." Probably the English equivalent is "Count ten before you speak."

1. RECKONING, NUMBER

The word logarithm is derived from two Greek words. "Logos" means "reckoning" and "arithmos" means "number." There is a distinct possibility that Napier took the word "logos" to mean "calculation" as in the English word "logistic." Hence "logarithm" can be said to mean "number calculation."

2. 2

The logarithm of a number is the power to which the base must be raised to equal the number, or writing this in the form of an equation:

$$NUMBER = (BASE)^{LOGARITHM}$$

But we know that $16 = 4^2$
which means that the logarithm of 16 to the base 4 is 2.
This expression is written $\log_4 16 = 2$.

3. YES

Isaac Newton was born on the 25th of December 1642 and he died on the 20th of March 1727. In 1614 a book in Latin was published in which the discovery of logarithms was recorded. An English translation of this book appeared in 1615 and tables of logarithms to the base e were published in 1619. Within a year tables of common logarithms were also published.

4. NAPIER AND BÜRGI

The Scots would recognize John Napier, Baron of Merchiston, as the first to discover or invent logarithms. The Swiss would be equally emphatic that Joost Bürgi, a Swiss astronomer and a friend of Kepler, should be given the honor of being the inventor. There is little doubt that the idea behind this wonderful aid to calculation was made quite independently by these two mathematicians early in the seventeenth century.

5. BRIGGS

Henry Briggs was educated at Cambridge but became professor of mathematics at Oxford in 1620. He said with reference to Napier's book introducing logarithms that he never saw a book which pleased him more. Briggs could see that if the base of logarithms were 10 and not e, then a greater simplification would result and the new logarithms would be of greater practical value. He set to work and published in 1624 his *Arithmetica Logarithmica*, giving the common logarithms of 30,000 numbers. These logarithms were extended to fourteen places of decimals!

6. DUTCH

Adrian Vlacq of Leyden supplemented Briggs' table of common logarithms. Apparently Briggs had included in his tables only logarithms from 1 to 20,000 and from 90,000 to 100,000. Vlacq filled in the gap between these numbers, so that the 1628 table contained the logarithms of all the numbers from 1 to 100,000.

7. NAPIER

The full name of the *Descriptio* is *Mirifici Logarithmorum Canonis Descriptio*. This book was published three years before the death of John Napier, in 1617. It contained his table of logarithms with the rules for its use, but there was no account of the construction of the table. The construction was explained in a work published after Napier had died. It is interesting to know that Briggs consulted Napier about the compilation of the new tables to the base 10, and that Napier had already thought of the practical possibility of such a table.

8. MULTIPLY BY $LOG_{10}e$

Natural logarithms may be converted into common logarithms by multiplying by a factor called the "modulus" of the common system of logarithms. This "modulus" is $\log_{10}e$

or $\dfrac{1}{\log_e 10}$ which has a value of 0.43429448 . . .

9. SLIDE RULE

This is a mathematical instrument used for calculations. The figures on each part of the instrument are spaced out in length according to the logarithms of their values. The modern slide rule consists essentially of two scales, which are marked on pieces of wood, one of which can slide along the other. The process of multiplication is thereby reduced to one of addition of lengths, and this is easily done by means of the sliding scales. The modern instrument can be used for many different types of calculations, and is much more complicated than the original design by William Oughtred.

10. GREATEST: LOG 2 + LOG 4,
LEAST: LOG 6 − LOG 3

$\log(2 + 4) = \log 6$
$\log 2 + \log 4 = \log(2 \times 4) = \log 8$
$\log(6 - 3) = \log 3$
$\log 6 - \log 3 = \log(6 \div 3) = \log 2$
Log 8 is the greatest and log 2 is the smallest of these values.

1. *RIGHT*
The answer is not correct according to Charlie's way of thinking—merely to reverse all the numbers is not the correct way.

2. *RIGHT*
Again it is the right answer but the wrong way. Will it happen once more?

3. *RIGHT*
$$\sqrt{5\tfrac{5}{24}} = \sqrt{\tfrac{125}{24}} = \sqrt{\tfrac{25\cdot5}{24}} = 5\sqrt{\tfrac{5}{24}}$$

4. *RIGHT*
$$\sqrt[3]{2\tfrac{2}{7}} = \sqrt[3]{\tfrac{16}{7}} = \sqrt[3]{\tfrac{8\cdot2}{7}} = 2\sqrt[3]{\tfrac{2}{7}}$$

5. *RIGHT*
Charlie is correct this time but does he realize why this is so? His conclusion does not necessarily follow from the first part of the question. He does not use the word "therefore" correctly.

6. *RIGHT*
He succeeds in obtaining the answer, but $\sin(a + b)$ does not equal $\sin a + \sin b$! In fact $\sin(a + b)\cdot\sin(a - b)$ should be $\sin^2 a\cdot\cos^2 b - \cos^2 a\cdot\sin^2 b$, and the solution follows from here.

7. *RIGHT*
You will notice that you are given two equations to find two unknowns, and Charlie uses only one of them. Also he writes in error that $\dfrac{x - 1}{y} = \dfrac{x}{y} - 1$. Finally, although $\dfrac{x}{y} = \dfrac{5}{6}$, it does not follow that $x = 5$, and $y = 6$; there are many other possibilities!

8. *RIGHT*
No working is necessary here—just count the number of triangles.

Although Charlie's work shows consistently correct answers, there is an amazingly incorrect use of "therefore" in writing and by symbol! In the first four examples, it is essential to show the working. Perhaps he has worked his examples properly on a piece of scrap paper, or perhaps he has just guessed.

1. 120

This is an illustration of the multiplicative principle. In this case there are twelve ways of choosing the boy for president. With each of these ways it is possible to choose the girl for vice-president in ten ways. The particular girl who is chosen is not determined by the choice of the boy. The choice of each is made independently and in succession, so that the total number of possibilities is the product of the two possibilities.

2. 720

This may seem a big number of arrangements. It is the product of $6 \times 5 \times 4 \times 3 \times 2 \times 1$. Another way of writing this product is ⌊6, or, as it is often printed, 6!. It is called factorial 6. In this example, the left-hand boy can be any one of them, so there are six ways of choosing him. The next boy from the left-hand side can be chosen in five ways from the remaining five boys. The next boy in four ways, the next boy in three ways, and so on. If there were eight boys altogether (only two more), the number of possible arrangements would be 8! or 40,320. If there were ten boys, then there would be more than three million ways of arranging them.

3. 120

This is not the same answer as in the last question because it is only the order which is considered and not the actual position. There will be six positions in which the same order will be found but each position will be turned around relatively to the other. Another way of considering this problem is to keep one boy always in the same place and then arrange the remaining five boys. This can be done in 5! ways, or 120. Any order arranged clockwise has an equivalent order arranged counterclockwise. The number of 120 different ways includes both these as separate arrangements. It is considered that sitting on a person's right is different from sitting on his left. These two arrangements are mirror images of one another.

4. $26 \times 25 \times 24 = 15,600$

This is called a "permutation of 26 different letters taken 3 at a time" and is written in mathematical language as $_{26}P_3$. It is fairly easy to see how one arrives at this calculation. Expressed in terms of factorials, it is the result of dividing factorial 26 by factorial $(26 - 3)$. In general the number of permutations of n things if only r are taken at any one time is $_nP_r$ or factorial n divided by factorial $(n - r)$.

5. $3 \times 3 \times 3 \times 3 = 3^4 = 81$

Consider each game separately. The first game may be won, lost, or drawn by one of the teams. Therefore there are three possibilities in this result. For each one of these first-game possibilities the second game has three possibilities. This makes nine possible forecasts for the first two games. For each of these nine forecasts the third game has three possibilities and so on. Hence for four games there are 81 different forecasts possible. If you are absolutely certain of the result of one of these games, then you need only make $3 \times 3 \times 3$ or 27 forecasts to ensure that among them is one complete correct forecast. If you can "bank" on two results, then you need only make 3×3 or 9 forecasts to ensure you have one correct forecast of all the four games.

6. 36

The argument is the same as in the last question. The first die may fall in six different ways and with each of these ways there are six possibilities for the second die. The total scores range from 2 to 12.

7. 1,728

What difficulty the captain will have in deciding the order of rowing in the boat for his crew if he has so many possibilities! This number 1,728 can be obtained in several ways.

Consider the stroke-side men first: the fourth oarsman can be chosen from the three who can row on either side in three ways; when this fourth oarsman is chosen, the four stroke-side oarsmen can be arranged in 4! ways; therefore there are 3 × 4! ways of arranging the stroke side. Now consider the bow-side oarsmen. There is no choice of men. There are two bow-side oarsmen and two who can row on either side. These can be arranged in 4! ways. For each stroke-side arrangement any one of the bow-side arrangements is possible. Thus the total number of arrangements is 3 × 4! × 4!, which is 1,728.

1. 39 DAYS
During the last and fortieth day the pond which was half covered becomes completely covered—just doubled in one day!

2. 2 GALLONS
A volume has three dimensions and each is doubled according to the question. Hence the new volume is $2 \times 2 \times 2$ times the original volume.

3. PHILADELPHIA
The train leaving Atlantic City travels the faster, so naturally they meet and cross one another nearer to Philadelphia. In fact, the meeting place is $^{40}/_{90}$ of 60, or $26\frac{2}{3}$ miles from Philadelphia, and $^{50}/_{90}$ of 60, or $33\frac{1}{3}$ miles from Atlantic City, and this happens at 10:40 A.M.

4. (a) 15, (b) 7, (c) 21
(a) The numbers in this series double themselves for each new term.
(b) These numbers are the cubes of the natural numbers.
(c) Each number in this series is five greater than the previous number.

5. (a) 9, (b) 16, (c) 20
(a) Each term in this series is one third of the previous term.
(b) These numbers are the squares of the natural numbers.
(c) These numbers are 1×2, 2×3, 3×4, 4×5, 5×6, . . .

6. SIX
The centers of the surrounding coins lie on a circle of radius equal to the diameter of the coin. The centers form the corners of a regular hexagon.

7. I, II, III, IIII, V, VI
The usual way of writing "four" in Roman numerals is IV.

[150]

ANSWERS TO QUIZ NO. **39**

1. *SPAIN*
One peseta = 100 centimos. Notes and coins are available. The rate of exchange is about forty pesetas to the dollar.

2. *THAILAND or SIAM*
One baht = 100 satang. The baht was previously called a tical. It is worth about four and one half cents.

3. *AUSTRIA*
One schilling = 100 groschen. The tourist counts a schilling as four cents, for one dollar is worth about twenty-five schillings.

4. *GREECE*
One drachma = 100 lepta, and about thirty drachmae equal one dollar.

5. *JAPAN*
One yen = 100 sen, and one dollar is worth 350 yen.

6. *PORTUGAL*
One escudo = 100 centavos. Each escudo is worth about four cents, for twenty-eight escudos will buy one dollar.

7. *ITALY*
One dollar is approximately equivalent to 600 lire.

8. *PERU*
One sol = 100 centavos. Nineteen soles equal about one dollar.

9. *BURMA*
One kyat is worth about twenty-one cents, so that the cost per day is about three dollars.

10. *BULGARIA*
One leva = 100 stotinki. One dollar is worth seven levas.

[151]

1. *d.* This is derived from the Roman word "denarius." British money had its origin in Roman times when 1 libra (pound) = 20 solidi, and 1 solidus = 12 denarii. Hence the British abbreviations L. S. D.

2. *n.* This product is known as "factorial *n*" or "*n* factorial." It is sometimes written as $\lfloor n$, but *n*! is easier to print.

3. *a.* As in the alphabet so in both arithmetical progression and geometrical progression, *a* is the first term.

4. *a.* This stands for area, which in the case of a rectangle equals the product of the length (*l*) and the base (*b*).

5. *c.* This means circumference, for it is π times the diameter of a circle.

6. *s.* This is usually written as a small "*s*." The capital "*S*" is reserved for one half of the sum of the angles of a spherical triangle.

7. *i.* Any number of the form $a + bi$ where $b \neq 0$ and $i = \sqrt{-1}$ is an "imaginary number."

8. *r.* These numbers form a geometrical progression whose common ratio is 3. The letter used to represent common ratio is "*r*."

9. *l.* This is the abbreviation for 1,000 cubic centimeters or 1 liter.

Hence the letters form the anagram DNAACSIRL which when rearranged give the word CARDINALS.

1. 1 6 15 20 15 6 1
These numbers are obtained by adding together the figures
found on the left hand and the right hand immediately above
the dashes.

2. 9 36 84 126 126 84 36 9
The extreme numbers on either side of the line are both 1.
The other numbers are obtained as explained in the answer
to question number 1. In exactly the same way, line after line
can be added indefinitely to the triangle.

3. *THIRD AND FOURTH LINES*
The full expansions are:
$$(x + a)^2 = 1(x^2) + 2(ax) + 1(a^2)$$
$$(x + a)^3 = 1(x^3) + 3(ax^2) + 3(a^2x) + 1(a^3)$$
The numbers or the coefficients of the terms in this type of
expansion are readily obtained by the direct application of
the binomial theorem which states that:

$$(x + a)^n = x^n + nx^{n-1} a + \frac{n(n - 1)}{1 \cdot 2} x^{n-2} a^2 +$$

$$\frac{n(n - 1)(n - 2)}{1 \cdot 2 \cdot 3} x^{n-3} a^3 + \text{etc., etc.}$$

4. 1 8 24 32 16
The numbers in the fifth line of the triangle are 1, 4, 6, 4, 1.
$\therefore$ the expansion of $(x + 2)^4 = x^4 + 4(x^3 2) + 6(x^2 2^2) +$
$4(x 2^3) + 2^4$. From this the coefficients as above are derived.
Both the coefficients and the actual terms are found by sub-
stituting $a = 2$ and $n = 4$ in the binomial expansion stated
in the last answer thus:

$$(x + 2)^4 = x^4 + 4x^3 2 + \frac{4 \cdot 3}{1 \cdot 2} x^2 2^2 + \frac{4 \cdot 3 \cdot 2}{1 \cdot 2 \cdot 3} x 2^3 +$$

$$\frac{4 \cdot 3 \cdot 2 \cdot 1}{1 \cdot 2 \cdot 3 \cdot 4} x^0 2^4$$

$$\therefore (x + 2)^4 = x^4 + 8x^3 + 24x^2 + 32x + 16$$

[153]

5. PASCAL

Pascal contributed much to mathematics, including a paper on the arithmetical triangle. Hence the name "Pascal's triangle." Other mathematicians—Tartaglia (1560), Schenbel (1558), and Bienewitz (1524)—had used this to determine the coefficients in a binomial expansion, but they had not dealt with the arrangement as thoroughly as Pascal did in his *Traité du triangle arithmétique,* printed in 1654.

6. 15. MAGIC SQUARE

A magic square is an arrangement of numbers such that the sum in each row, each column, and each diagonal is the same. The earliest magic square dates back to about 2200 B.C., and no doubt it was familiar to the Chinese then. The squares were arranged in "orders," and one can appreciate why magic properties, such as longevity and disease prevention, became associated with these squares.

7.	16	2	12
		6	10	14
	8	18	4

To obtain the missing numbers, first find the total of the top row or the left-hand column. This equals 30. Then calculate the center number from the diagonal. The rest follow easily from this point.

8.	1	15	14	4
	12	6	7	9
	8	10	11	5
	13	3	2	16

This is a magic square using all the numbers 1 to 16 only. It is formed from a basic square in which all these numbers are included written in order left to right across each row in turn. All the numbers cut by both diagonals are retained but all the others are interchanged with their diametrically opposite numbers.

[154]

1. 1, 1·2, 1·2·3, 1·2·3·4, 1·2·3·4·5, ...

Each term in this series is a factorial, which means the product of all the numbers from 1 to the particular term considered. The first five terms of this series are thus: 1, 2, 6, 24, 120, and the sum of these is 153.

2. *HERE THEY ARE:*

$$\frac{2}{4} = \frac{3}{6} = \frac{79}{158}, \qquad \frac{3}{6} = \frac{9}{18} = \frac{27}{54}, \qquad \frac{2}{6} = \frac{3}{9} = \frac{58}{174}$$

3. 1 *POUND*, 3 *POUNDS*, 9 *POUNDS*

The key to this is that the boy can put any combination of weights on either pan and the difference between the two weights is the amount of fruit he sells. Thus the weight he requires is the result of an addition or subtraction sum.

$1 - 0 = 1$, $3 - 1 = 2$, $3 - 0 = 3$, $(3 + 1) - 0 = 4$, $9 - (3 + 1) = 5$, $9 - 3 = 6$, $(9 + 1) - 3 = 7$, $9 - 1 = 8$, $9 - 0 = 9$, $(9 + 1) - 0 = 10$, $(9 + 3) - 1 = 11$, $(9 + 3) - 0 = 12$, and finally $(9 + 3 + 1) - 0 = 13$.

4. *YES*

$$\frac{9}{\sqrt{9} \times \sqrt{9}}, \qquad \frac{9}{9} + \sqrt{9}, \qquad \frac{9}{\sqrt{9}} + \sqrt{9}$$

5. *NORTH POLE*

Yes, and there are other places too! These places lie on a parallel of latitude in the Southern Hemisphere 100 miles north of a parallel of latitude that has a total length of 100 miles. Can you puzzle this out?

6. 264 feet, 880 yards

> 60 miles per hour = 88 feet per second
>
> ∴ length of train = 88 × 3 feet
> = 264 feet

To pass completely through the tunnel, the train must travel for a time of thirty seconds.

> ∴ length of tunnel = 88 × 30 feet
> = 880 yards

1. *THE SIGN FOR EQUALITY*
There is little doubt that the parallel lines of one length, =, which is the sign for equality today, was due to Robert Recorde. In his book *The Whetstone of Witte*, written in 1557, he clearly states why he introduced the symbol. Other symbols did exist until the eighteenth century, but we must admit that "no two things can be more equal" than these two lines.

2. π
In the sixteenth century there appeared to be an urge by mathematicians to calculate the value of π to many places of decimals. Ludolph (or Ludolf) van Ceulen gave the value of π to some twenty places of decimals in 1596. A few years later, he had worked out the ratio to thirty-five places. Hence the reason why π was called in German textbooks the "Ludolphische zahl."

3. *ALGEBRA*
Toward the close of the fifteenth century, Lucas Pacioli, a Franciscan friar, published a mathematics book in Venice. Like the Arabs before him, he called the unknown quantity the "thing" (we of course use the letter "x"). The Italian word for "thing" is "cosa." Hence we see why the old name for algebra in England was the "cossic art" or the "rule of coss."

4. *PRIME NUMBERS*
Eratosthenes was a contemporary of Archimedes and both were educated at Alexandria. He constructed an instrument to duplicate a cube and gave a laborious method of constructing a table of prime numbers. The latter is called the "sieve of Eratosthenes," and as one bright pupil said, "If any composite number gets through that sieve then the age of miracles is not passed."

5. A PARTICULAR DIVISION OF A LINE

If you draw any line AB and then find a point C in it such that $\dfrac{AB}{AC} = \dfrac{AC}{CB}$ then the line AB is cut in the golden section at the point C. This section is used in art and it is pleasing to the eye to see a composition on canvas where the golden section has been intentionally or intuitively applied.

6. MORE THAN ONE POSSIBLE SOLUTION FOR A TRIANGLE

If you are given two sides of a triangle and an angle opposite one of them, it is very likely that two triangles can be found that will fulfill these conditions. For instance, if $a = 100$ feet, $b = 224$ feet, and angle $A = 30°$, then if you solve the triangle one solution is possible where angle B is less than $90°$ and another where angle B is greater than $90°$.

7. SUNDIAL

The style is the pin, rod, or triangular plate which forms the gnomon of a sundial. A pupil of Thales, the founder of the first Greek school of mathematics, is said to have introduced the use of the style or gnomon into Greece. Simple sundials have been found in Pompeii.

8. CHECKING MULTIPLICATION AND ADDITION

This method of "casting out the nines" was introduced nearly a thousand years ago by the Arabs. Nines are "cast out" of each factor in the multiplication equation. The remainders are then multiplied and nines are "cast out" again. If the remainders at this stage are unequal, the equation is false. It does not follow that if the remainders are equal the equation is true!

Example: Let us find out if this multiplication is false.

$$7{,}926 \times 3{,}487 = 27{,}637{,}862.$$

"Cast out" the nines and the remainders are:

6   4   5

Multiply the remainders:

24   5

"Cast out" the nines again and the remainders are:

6   5

These remainders are unequal; therefore there is a mistake in the equation. Actually, the answer should be 27,637,962. If this process of "casting out the nines" had been applied in the same way to the product 27,637,692, then the error would not have been detected. Thus the process has only a limited application.

The same process of "casting out the nines" can be used to check for mistakes in the addition of numbers.

9. ARC OF A CYCLOID

The question refers to the well-known brachistochrone problem or the curve on which a body descending to a given point under the action of gravity will reach it in the shortest time. The Bernoullis are responsible for the name of the problem. At first sight, it does seem amazing that a steel ball will roll from one point to a lower point on a cycloidal-shaped bowl in a quicker time than it will roll down a plane joining the two points. Refer to the diagram—the steel ball
will roll down from A to B more
quickly along path C than it will
along path P.

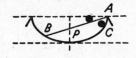

10. JAPANESE ABACUS

The name of the instrument is sometimes spelled with an "o," thus: soroban. It is very similar to the Chinese abacus known as a suanpan or swanpan. There is no doubt that the Japanese can use their abacus with amazing speed and dexterity. The most complicated sums are worked out with unerring accuracy. The Japanese shopkeepers and clerks become very dependent upon their sorabans, for they are used frequently in sums which should be worked out mentally.

ANSWERS TO QUIZ NO. 44

1. *HARMONIC PROGRESSION*

This is sometimes abbreviated to H.P., but don't confuse it with a horse! If a, b, and c are in harmonic progression, then $\frac{1}{a}$, $\frac{1}{b}$, and $\frac{1}{c}$ are in arithmetical progression. Thus 1, $\frac{1}{2}$, $\frac{1}{3}$, $\frac{1}{4}$ ... form a harmonic progression, for 1, 2, 3, 4 ... are in arithmetical progression. In music strings of the same material with the same diameter and tension, but with lengths in harmonic progression, produce harmonic tones.

2. *FIBONACCI SERIES*

This series consists of the numbers 0, 1, 1, 2, 3, 5, 8, 13, 21, ... The sum of the second and third terms equals the fourth term, and the sum of the third and fourth terms equals the fifth term, and so on all the way through the series. This series is named after Fibonacci, who is generally known as Leonardo of Pisa. He was born in 1175. Curiously, the leaves on a stalk, the petals of some flowers, and lettuce leaves follow the ratios of successive terms of the Fibonacci series.

3. *LET US SEE!*

$$S = 1 + 3x + 5x^2 + \ldots + 39x^{19}$$
$$Sx = x + 3x^2 + \ldots + 37x^{19} + 39x^{20}$$

By subtraction:

$$S(1-x) = 1 + 2x + 2x^2 + \ldots + 2x^{19} - 39x^{20}$$
$$= 1 + \frac{2x(1-x^{19})}{1-x} - 39x^{20}$$
$$= \frac{1 - x + 2x - 2x^{20} - 39x^{20} + 39x^{21}}{1-x}$$
$$\text{Sum} = \frac{39x^{21} - 41x^{20} + x + 1}{(1-x)^2}$$

[159]

4. (a) $19 \cdot 21 \cdot 23$ (b) 28,560

In order to find the sum of n terms of this series, write down the last term multiplied by the next highest factor and subtract the first term multiplied by the next lowest factor, and then divide by the product of (the number of factors in each term + 1) and (the common difference of the factors). Obeying these instructions in this particular series we get:

$$\text{Sum} = \frac{19 \cdot 21 \cdot 23 \cdot 25 - 3 \cdot 5 \cdot 7 \cdot 9}{(3+1) \cdot 2}$$

$$= \frac{229{,}425 - 945}{8}$$

$$= 28{,}560$$

5. *NO*

At first sight this may appear strange, but if we put $x = 1$ in this series we obtain:

$$\log_e 2 = 1 - \frac{1}{2} + \frac{1}{3} - \frac{1}{4} + \frac{1}{5} - \ldots \ldots \ldots$$

In order to get the value of the logarithm correct to four places of decimals, we should have to consider hundreds of terms of this series. Life is too short for this! Fortunately other series can be deduced from $\log_e (1 + x)$ which are more quickly convergent and hence more useful for the calculation of logarithms.

6. *Sin x*

Both sin x and cos x are functions of x and they can both be expanded in a series of terms which are in ascending powers of x. Using Maclaurin's theorem, it is a simple matter to show that

$$\sin x = x - \frac{x^3}{3!} + \frac{x^5}{5!} - \frac{x^7}{7!} + \ldots \ldots \ldots$$

[160]

x in these expansions is measured in radians and not degrees. (π radians $= 180°$). Using this series we can soon work out the value of various sines. For instance, if we put $x = 0.2$ in the series we shall find that sin x has a value of 0.1988. Now convert x from radian measure into degrees:

$$x = \frac{0.2 \times 180}{3.14} \text{ degrees}$$
$$= 11° 28'$$

Thus we succeed in calculating that sin $11° 28' = 0.1988$.

7. EXPONENTIAL

The base of natural logarithms is e, and this quantity is the sum to infinity of the series:

$$1 + \frac{1}{1!} + \frac{1}{2!} + \frac{1}{3!} + \frac{1}{4!} + \cdots\cdots\cdots\cdots$$

The peculiarity of this series is that when it is raised to the power x the following series is obtained:

$$1 + \frac{x}{1!} + \frac{x^2}{2!} + \frac{x^3}{3!} + \frac{x^4}{4!} + \cdots\cdots\cdots\cdots$$

and this series is convergent for all finite values of x. This series, which is the expansion of e^x, is known as the exponential series.

ANSWERS TO QUIZ NO. 45

1. THALES
This Greek philosopher lived about 600 B.C. He introduced abstract geometry, and in the process of trying to develop deductive reasoning he collected a number of geometrical facts together. Many discoveries in geometry are attributed to Thales. He differed from some of the Greek mathematicians who came after him because he applied geometry to a number of practical problems.

2. EUCLID
He was born about 330 B.C., and was of Greek descent. Euclid is without doubt the most famous of all mathematicians before Newton. He was the author of several works, but one of the world's famous books, the *Elements,* made his reputation. In the thirteen books, Euclid collected all the known facts about geometry, and it was at once accepted as the standard work. It was translated into Arabic, and came back to Europe in a Latin translation.

3. APOLLONIUS
He lived from about 260 to 200 B.C. He studied at Alexandria and probably lectured there. Later, he spent some years at Pergamum before returning finally to Alexandria. He is celebrated for a systematic work on conic sections. Apollonius is called the great geometrician. There was evidently no Greek rule that constructions should be done with a compass and a ruler only, for Apollonius was the first person who definitely stated this requirement.

4. LEIBNITZ
Gottfried Wilhelm Leibnitz (or Leibniz) was most catholic in his studies and interests. He is best known for his discovery in the seventeenth century of calculus, which he did independently of Newton. Leibnitz used the names "calculus differentialis" and "calculus integralis," so there is no doubt who was the Father of the names in common use today. He is also responsible for the notation of calculus.

[162]

5. GAUSS

Carl Friedrich Gauss was born in 1777, and became a great astronomer as well as a brilliant mathematician. His work *Disquisitiones arithmeticae* was published in 1801, and it is still one of the standard works on the theory of numbers. From a magnetic observatory erected at Göttingen, he sent telegraphic signals to a neighboring town, and was the first to show the practicability of the electromagnetic telegraph.

6. RECORDE

Robert Recorde, who studied at Oxford and graduated in medicine at Cambridge in 1545, was the author of *The Grounde of Artes*. This is an arithmetic book and it is one of the earliest mathematical books to be printed in English. The book, published in 1540, "teaches the work and practice of arithmetic in whole numbers and fractions," and is in the form of a catechism or dialogue. He used the + and the − signs to represent "too much" and "too little" respectively.

7. BERNOULLI

One of the famous family of Swiss mathematicians was James (Jacob or Jacques) Bernoulli, who was appointed as professor of mathematics at Basle in 1687. He did much to extend the use of calculus, and he was the first mathematician to publish a work on integral calculus. In his *Ars Conjectandi* he defines the numbers which are now named after him. They are the numerical values of the coefficients of

$$\frac{x^2}{2!}, \frac{x^4}{4!}, \frac{x^6}{6!}, \ldots \ldots \ldots \frac{x^{2n}}{(2n)!}, \text{ in the expansion of } \frac{x\,e^x}{(e^x - 1)}$$

8. LAGRANGE

He is said to have been the greatest mathematician of the eighteenth century. He was born in Turin and established an academy there. He later went to Berlin and finally to

Paris as a professor. He made original contributions to mathematics at a very early age, and he was a ceaseless writer all his life. His monumental work is the *Mécanique analytique,* in which he deduces the whole of mechanics from one fundamental principle. It was in an earlier work while at Turin that he dealt with the problem of the transverse vibration of stretched strings, and he pointed out the deficiencies in earlier solutions and gave the complete solution.

9. MERCATOR

He was a Flemish mathematician and geographer who devoted his life to mathematical geography. His great world map was completed in 1569. It is called Mercator's projection and has the parallels and meridians at right angles. Compass bearings can be drawn as straight lines on these maps, and since the eighteenth century this projection has been used for nautical charts. Mercator made maps, globes, and astronomical instruments of wonderful precision considering that they were constructed many years ago. His work was carried on by two of his sons.

10. AHMES

The so-called Rhind papyrus, which is now in the British Museum, was written by a scribe named Ahmes. It is one of the earliest mathematical documents on papyrus existing, and dates back some fifteen hundred or more years B.C. It consists of some rules and questions in arithmetic and geometry. The answers are given but not the working. Quite a lot of mensuration is to be found on the Rhind papyrus. It has this name because it was purchased, more than a hundred years ago, by an English Egyptologist named Rhind. Ahmes is clearly copying an older work, for he writes, "This book was copied in the year 33," and the work is headed, "Directions for knowing all dark things."

ANSWERS TO QUIZ NO. 46

ACROSS

1. *ADDITIONS.* "Tot" is the short for "total" or the Latin "totum." It appears to have come into use about 1690.

2. *TRAP*

3. *STAR.* The pointers in the constellation Ursa Major or "The Big Dipper" point to the North Star.

4. *RL.* Reversed abbreviation of latus rectum, or the chord perpendicular to the axis of the parabola and passing through its focus.

5. *NH*

6. *LV*

7. *APPLE.* This recalls the well-known story of the apple falling from the tree to the ground.

8. *CR.* Abbreviation for credit and typed in black on a bank statement. You are not then "in the red."

9. *IN.* Isaac Newton. Through his work on gravitation, he discovered the laws governing the movements of the planets around the sun, the action of the moon on the tides, and how to predict the courses of comets.

10. *LB.* The area equals length times breadth, or, written in symbols, $A = l \cdot b \cdot$

11. *CM.* The abbreviation for centimeter.

12. *HP.* Horsepower, which is the unit of the rate of doing work.

13. *MEASURING*

[165]

DOWN

3. SNELL. Dutch astronomer and mathematician who was born in Leyden in 1591, where he later became professor of mathematics. He discovered the law of refraction in 1621.

14. DINAR

15. INS. The correct abbreviation for inches is "in."

16. IOU. This represents "I owe you" and is the formal acknowledgement of a debt.

17. NORTH. When the north direction is marked, the plan can readily be orientated.

18. AUTOLYCUS. He flourished about 330 B.C. Two Greek works on astronomy exist and are preserved at Oxford.

19. BAR GRAPHS. These graphs consist of parallel bars whose lengths are proportional to certain quantities given in a set of tables.

20. PLANE

21. ROPES. In order to mark out a right angle the Egyptians used knotted ropes to form a "three-, four-, and five-sided" triangle.

22. ONE

23. CGS. The centimeter–gram–second system, where these are the fundamental units of length, mass, and time.

24. BAR. In order to show a negative characteristic in a logarithm the minus sign is written over the figure.

25. SIN. This is the abbreviation for the sine of an angle.

1. HISTOGRAM

Let us suppose that we know the heights of a thousand men picked at random. Tabulate the number of men whose heights lie in the equal ranges between 64 and 65 inches, 65 and 66 inches, 66 and 67 inches, and so on. Plot on a graph the frequency (number of men) in each range against the heights by making a series of columns whose areas are proportional to the frequency. This is a histogram and the diagram above the questions is one such graph.

2. FREQUENCY POLYGON

On the histogram mark the mid-points of all the ranges of the variable quantity as shown in the same diagram. Join these mid-points by a jagged line. This is a frequency polygon. When the number of observations is increased considerably, this frequency polygon becomes a frequency curve. This is made possible by choosing smaller ranges of the variable quantity and at the same time having a large number of observations in each range.

3. NORMAL CURVE

Many variable quantities form this shape of curve as their frequency curve. Some examples are: heights of persons, sizes of shoes worn by people, and intelligence quotients. One of the characteristics of this normal curve is that it is always smooth and symmetrical.

4. MEAN

This is what the ordinary person means when he speaks of an average. It is the "average" of everyday life. It is obtained by adding together all the values of the variable quantity and then dividing this by the total number of these values. In this way we should find, for instance, the average consumption of milk per person per day, or the average life span of a horse. Do batting averages work out in this way? What about the batters not out?

[167]

5. MODE

This is an average often used in statistics. It is the most commonly occurring value in a series of observations of a variable quantity. In a symmetrical frequency curve (or normal curve), the mean and the mode are the same. In the lopsided curve or skew curve, they are different.

6. STANDARD DEVIATION

Sometimes the observed values of a variable quantity are all close to the mean value. Sometimes they are widely dispersed. The standard deviation tells us the degree to which they are dispersed. This standard deviation, denoted by the Greek letter sigma, σ, is calculated by taking the square root of the arithmetic mean of the squares of the deviations from the mean. Mean error, mean square error, and error of mean square are other names for standard deviation. σ^2 is called the variance.

7. RANDOM

The words "random sample" abbreviate the phrase "a sample chosen in a random way." This type of sampling is often carried out by making use of a table of random numbers. After numbering the items in a population, it is then a simple matter to select those items or samples by consulting the table. This is done to prevent personal bias. Such sampling is obviously necessary for a dealer buying apples, a metallurgist in testing material, a manufacturer of electric-light bulbs, etc. Random sampling saves time and energy.

1. *HALFWAY*
If you go farther you are coming out again and nearer the edge!

2. *EVERY FOUR MINUTES*
The buses are evenly spaced along the highway in both directions. He counts every hour twenty moving in one direction and ten in the opposite direction. There are thirty buses in that section over which he travels in an hour, and half of these buses turn around in the hour. Therefore fifteen must leave the terminus every hour or one bus every four minutes.

3. *TWELVE*

4. 400 π *FEET*
Subtract the circumference of the earth 2π (4,000 miles) from the circumference flown by the airplane 2π (4,000 miles + 200 feet). The difference is 2π 200 feet.

5. 20,000
Assuming also that one man has one wife, then:

$$\frac{42}{100} \text{ of the males} = \frac{28}{100} \text{ of the females}$$

$$\text{or} \quad 42M = 28F$$
$$\text{or} \quad M : F = 28 : 42$$
$$\text{or} \quad M : F = 2 : 3$$

Therefore two-fifths of the population is male, or there are 20,000 males.

6. *YOU WILL NOT BE ABLE TO HOP OUT*
You hop 4½ feet at the first attempt, which is halfway out, and then another 2¼ feet at the next hop. Thus you are

[169]

already three quarters of the way out in two hops. You feel encouraged, for surely the last quarter will be hopped easily! Let us write down the hops:

$$4\tfrac{1}{2}, \; 2\tfrac{1}{4}, \; 1\tfrac{1}{8}, \; \tfrac{9}{16}, \; \tfrac{9}{32}, \; \tfrac{9}{64}, \; \tfrac{9}{128}, \; \tfrac{9}{256}, \; \tfrac{9}{512}, \; \tfrac{9}{1024}, \text{ and so on.}$$

Add these up and you will see that you are nearly there—in fact, you can hop more than 8¾ feet of the total distance needed of 9 feet. But this is a series whose "sum to infinity" is less than 9 feet. You are a prisoner in the circle!

7. YES

The first box could contain $(x + 1)$ or $(x + 2)$ pieces of candy and the second box could contain $(y + 1)$ or $(y + 2)$ pieces, where both x and y separately are divisible by three. Then neither of the two boxes alone could contain a number of candies divisible by three.

There are four possible ways in which these boxes can be filled with candy, and they are:

$$
\begin{array}{lll}
(x + 1) & \text{and} & (y + 1) \\
(x + 2) & \text{and} & (y + 1) \\
(x + 2) & \text{and} & (y + 2) \\
(x + 1) & \text{and} & (y + 2)
\end{array}
$$

The other point noted by the student is that the second box has seven more pieces than the first box, which can be expressed in these four ways:

$(y + 1) - (x + 1) = 7$ or $y - x = 7$, which is impossible.
$(y + 1) - (x + 2) = 7$ or $y - x = 8$, which is impossible.
$(y + 2) - (x + 2) = 7$ or $y - x = 7$, which is impossible.
$(y + 2) - (x + 1) = 7$ or $y - x = 6$, which is possible because six is divisible by three.

Thus it is possible to fill the two boxes with candy in such a way that the conditions of the question are fulfilled. In

[170]

fact, if we remember that y is always six greater than x, and that x is always a multiple of three, then we can calculate pairs of values for $(x + 1)$ and $(y + 2)$.

Here are some ways that the two boxes can be filled with candy:

First box	. . . 1	and	second box	. . . 8
	. . . 4			. . . 11
	. . . 7			. . . 14
	. . . 10			. . . 17
	. . . 13			. . . 20

1. *DIRECTOR*

The locus of the intersection of pairs of perpendicular tangents to an ellipse is a circle, and this circle is called the director circle of the ellipse.

2. *AUXILIARY*

The circle which is described with the major axis of the ellipse as diameter is called the auxiliary circle of the ellipse. Unlike the director circle, the auxiliary circle will touch the ellipse in two points at the extremities of the longer or major axis. Draw a circle, and from a number of points on it drop perpendiculars on a diameter. Divide these perpendiculars in a given ratio (say 2:3). The join of these points will form an ellipse with the original circle as the auxiliary circle.

3. *NINE-POINT*

This circle as its name suggests passes through nine points. Six of these points are mentioned in the question and the remaining three are "the mid-points of the lines between the vertices and the common point of intersection of the altitudes."

4. *ORTHOGONAL*

Orthogonal means "right-angled; pertaining to or depending upon the use of right angles." If any two curves cut at right angles, they are said to intersect orthogonally. Such curves are of interest in many branches of applied mathematics. One point of interest about two circles cutting orthogonally is that the square of the distance between the centers is equal to the sum of the squares of their radii.

5. *INSCRIBED*

A circle is said to be inscribed in a polygon when each side is tangential to the circle. Consider the case of the simplest polygon—a triangle. The inscribed circle is obtained by bisecting the angles of the triangle. These bisectors pass

[172]

through a common point which is the center of the inscribed circle.

6. *GREAT*

Any circle on the surface of a sphere whose plane goes through the center of the sphere is called a great circle. If the earth be considered as a sphere of radius 3,960 miles, the great circles passing through the North and South Poles are called meridians of longitude.

7. *OSCULATING*

An osculating circle of a curve has three or more coincident points in common with the curve. This term appears to have come into use early in the eighteenth century. The radius of the osculating circle gives the radius of curvature of a curve. This is usually found by making use of a formula in calculus.

8. *ESCRIBED*

The interior bisector of one angle and the exterior bisectors of the other angles of a triangle are concurrent in a point which is equidistant from one side and the other sides produced of a triangle. Hence, a circle can be described to touch one side and the other sides produced. Such a circle is called an escribed circle of the triangle, and every triangle has three escribed circles.

9. *CIRCUMSCRIBED*

The meaning of "to circumscribe" is to describe a figure around another so as to touch it at points without cutting it. This is precisely what takes place with the circumscribed circle. To find the center of such a circle, bisect the sides of a triangle and erect perpendiculars. They are concurrent at the circumcenter. The radius R of the circumscribed circle of the triangle ABC is given by

$$R = \frac{a}{2 \sin A} = \frac{b}{2 \sin B} = \frac{c}{2 \sin C}$$

[173]

1. 1½ MILES

Suppose the hiker strikes the road at X, a distance of x miles from P. Let T be the total time taken by the hiker to reach the inn. It is the time that has to be a minimum to fulfill the conditions of the question.

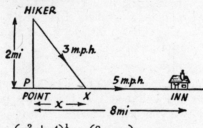

But $T = \dfrac{(x^2 + 4)^{\frac{1}{2}}}{3} + \dfrac{(8 - x)}{5}$

Differentiate T with respect to the variable x:

$$\frac{dT}{dx} = \frac{1}{3} \cdot \frac{1}{2} \cdot \frac{2x}{(x^2 + 4)^{\frac{1}{2}}} - \frac{1}{5}$$

There is a "minimum" when:

$$\frac{x}{3(x^2 + 4)^{\frac{1}{2}}} = \frac{1}{5}$$

or when $5x = 3(x^2 + 4)^{\frac{1}{2}}$
or when $x = 1\frac{1}{2}$ miles.

2. 96 π CUBIC INCHES

The section of the space hat is a parabola of the form $y^2 = 4ax$. Substitute one set of known values of the space hat and we have:

$$y^2 = 4ax$$
or $4^2 = 4a \cdot 12$

$$\therefore a = \frac{1}{3}$$

[174]

The equation of the parabola is $3y^2 = 4x$.

By integration we have the volume of the hat thus:

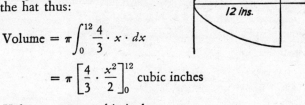

$$\text{Volume} = \pi \int_0^{12} \frac{4}{3} \cdot x \cdot dx$$

$$= \pi \left[\frac{4}{3} \cdot \frac{x^2}{2} \right]_0^{12} \text{ cubic inches}$$

$$\therefore \underline{\text{Volume} = 96\,\pi \text{ cubic inches}}$$

3. 1,152 *CUBIC INCHES*

Let V be the volume of the box.

Let x be the side of the squares cut out.

Then $V = (32 - 2x)(20 - 2x)x$

or $V = 640x - 104x^2 + 4x^3$

$\therefore \dfrac{dV}{dx} = 640 - 208x + 12x^2$

For a maximum volume, $\dfrac{dV}{dx} = 0$

or $3x^2 - 52x + 160 = 0$

or $(3x - 40)(x - 4) = 0$

or $x = 4$ or $x = \dfrac{40}{3}$

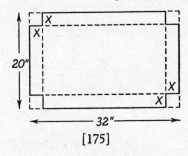

Using the real value of $x = 4$, the maximum volume of the box is $24 \times 12 \times 4 = 1,152$ cubic inches.

4. 573 FEET

Consider a small angular movement $d\theta$ takes place during a small displacement ds. Then if R is the radius of the circle in which it moves:

$$\frac{1}{R} = \frac{d\theta}{ds}$$

$$\therefore \; ds = R \, d\theta$$

$$\therefore \; s = R \int_0^{\frac{\pi}{6}} d\theta$$

$$= R \left[\theta \right]_0^{\frac{\pi}{6}}$$

$$\therefore \; 2 \cdot 150 = R \cdot \frac{\pi}{6}$$

$$\therefore \; \underline{R = 573 \text{ feet}}$$